Turning
BLOOD
The Fight for Life in Cooley's Anemia
Red

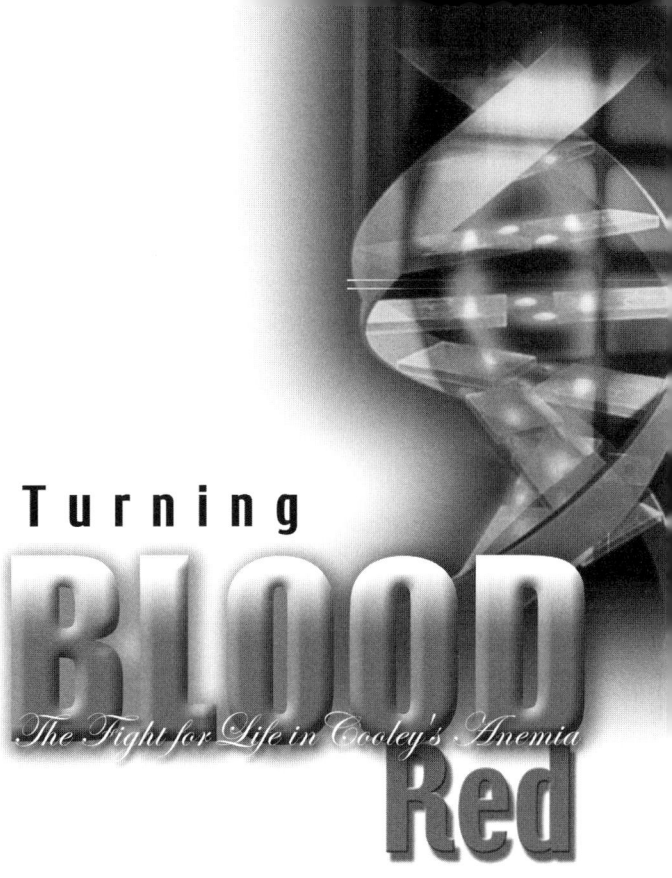

Turning
BLOOD
The Fight for Life in Cooley's Anemia
Red

Arthur Bank, M.D.

Columbia University, USA

 **World Scientific**

NEW JERSEY · LONDON · SINGAPORE · BEIJING · SHANGHAI · HONG KONG · TAIPEI · CHENNAI

Published by

World Scientific Publishing Co. Pte. Ltd.

5 Toh Tuck Link, Singapore 596224

USA office: 27 Warren Street, Suite 401-402, Hackensack, NJ 07601

UK office: 57 Shelton Street, Covent Garden, London WC2H 9HE

British Library Cataloguing-in-Publication Data
A catalogue record for this book is available from the British Library.

TURNING BLOOD RED
The Fight for Life in Cooley's Anemia

ISBN-13 978-981-283-247-4
ISBN-10 981-283-247-5

Typeset by Stallion Press
Email: enquiries@stallionpress.com

Printed by FuIsland Offset Printing (S) Pte Ltd, Singapore

This book is dedicated to patients with Cooley's anemia
and their families

Contents

Part 3: The Best Medicine: Current Care and Future Goals

Prologue

It was a snowy morning in 1966. The two little girls were bundled in parkas and boots and holding their mothers' hands tightly as they walked through the front door of the hospital, as usual teeming with people. It was Christmas week, but this was no vacation for either the two little children or their mothers. The children needed to be transfused with blood that day. It was vital. Otherwise, they would get sick and might die.

The two little girls and their mothers recognized each other in the lobby and took the same elevator. They all smiled at each other. The girls, still holding their mothers' hands, stepped into the crowded elevator which would take them to the sixth floor where they would receive their blood transfusions again. They saw other children they knew and their mothers get on the elevator, and they nodded hello. They knew the smells and closeness and pushing of the elevator well. They were there every two weeks.

In the transfusion clinic there were young patients waiting to be transfused; mothers helping them take off their jackets and sweaters; nurses setting up the intravenous sets; doctors writing orders and checking that the right blood was going to be given to the right patient; and scientists standing around in white coats hoping to get blood samples for their own experiments to find out more about this disease. It was a cauldron of interdependent human activity, blended together to treat a blood disease called thalassemia or Cooley's anemia.

What happened to those two little children holding their mothers' hands on the elevator? Where are they today? Did all of those patients being transfused end up the same way, whatever that way was? What is this disease about anyway? That is what this book will tell you.

This is a disease not well known to Americans. It is a story of tragedy, heartbreak, struggle, transcendence of the human spirit, and also courage and hope. Hope in the unflagging pursuit of greater knowledge and better care for patients today than was available to those two little girls long ago.

Chapter 1

Introduction

For over 40 years, as a scientist and clinician, I have been working on research to understand and cure a blood disease that kills thousands of children and young adults worldwide every year, and affects millions of people around the world. It is called Cooley's anemia, after Dr. Thomas Cooley, a physician who first described it in 1925. Children with the disease used to die in their teens and 20s. Now they live into their 30s and 40s and beyond as progress has been made in treatment. Although there is currently no reliable cure for the disease, there are realistic hopes and dreams that one will be discovered in the not too distant future.

This book is about Cooley's anemia: the patients; the families; the research to understand the disease; the ways to treat it; and the search for a cure.

Life is a struggle for all of us. We must survive disease, disappointment, adversity in the world around us, and the cruel blows of fate. This book is about people with a terrible blood disease that haunts their whole life and the struggle that entails. It is also the struggle of physicians caring for them, as well as the struggle of scientists trying to find a cure for the disease.

Linda De Pasquale has lived with Cooley's anemia for 45 years. She is a rare long-term survivor of the disease. But to keep herself alive and well, she has to stick herself with a needle and bear its discomfort for eight to twelve hours almost every day to receive the life-saving drug she needs. She also has to receive blood transfusions every two or three weeks.

After so many years with a disease in which bone erosion and cardiac complications are common, who knows when her bones will break or when her heart will truly fail? But for now, she is well. Her body and spirit survive, and she embraces every day as "the newest day" of the rest of her life. She enjoys her daily life on the combined fuels of medicine, friendship, love, and belief. What does she believe in? Herself, God, Buddha, her doctors, her friends. She believes in them all and is committed to life.

Who can tell her anything about her future that she doesn't already know? Looking out her window onto the waters off Sandy Hook, New Jersey, living her life as fully as she can, she says: "We're all going to die. It's simply a matter of when and how and how much we will suffer.

"We'll all get either cancer or heart disease or some other bad medical condition sometime. Why worry? Worrying and complaining are wasted emotions; they don't help. I just live life as it comes."

How has Linda coped? Her mother was her savior during childhood. When she died, her aunt, her doctors, and her friends took her place. But it is Linda D., with her commitment to complying with her treatment and her mental toughness, who has willed herself to survive so much longer than most other patients with her disease. She knows that, "If I don't take care of myself, no one else can help me.

"And someday there'll be a cure. But right now, I'm doing everything I can."

*   *   *

Cooley's anemia is a disease of the blood protein, hemoglobin. Hemoglobin transports oxygen in the red blood cells that circulate in our blood. There are billions of red cells circulating in our blood every minute of our lives. And each red cell has millions of hemoglobin molecules in it to transport oxygen to all of our cells.

Cells need oxygen for their power. Oxygen allows cells to make proteins, to turn fats and sugars into energy, and to keep the organs of our body functioning properly. Without oxygen, we die; without hemoglobin, we die.

Hemoglobin: The magic gift

Our red blood cells are simply porous bags of hemoglobin. Oxygen is transported through the body bound to hemoglobin. When oxygen in our lungs moves into our red cells, it is grasped by hemoglobin and then released to all of our body's cells as it is carried in our circulating blood. The blood cells are red because that is the color of hemoglobin bound to oxygen (oxygenated hemoglobin).

Hemoglobin has the unique and extremely efficient ability, developed over millennia, to pick up the oxygen we breathe as blood passes through our lungs. Just as efficiently, hemoglobin releases the oxygen as blood flows to all of our organs such as the kidneys, liver, brain, and heart.

Like a biological magic sponge, hemoglobin picks up oxygen as we breathe, and bathes our cells with it where it is needed for the cells to function well and stay alive.

Red cells and anemia

Red blood cells, like all other blood cells, are made in the bone marrow cavities inside our bones. The earliest red blood cells somehow know that their main job is to make large amounts of hemoglobin. We don't know how they know, but they know.

The earliest red blood cells made in our bone marrow have nuclei that carry our chromosomes and DNA. After dividing a few times, these nucleated red blood cells in the bone marrow, bursting with hemoglobin, become smaller, lose their nuclei, and leave the marrow to enter our circulating blood. Once in our blood stream, they become mature red blood cells that float around and live for months, providing us with the oxygen-carrying hemoglobin we need.

If there are too few red blood cells in our blood, or too little hemoglobin in each of the red blood cells, we have anemia. In either case, we have too little total hemoglobin and not enough oxygen delivery to our tissues.

A little anemia is tolerable; too much anemia is not. If the anemia is severe enough, we feel weak because our cells do not get enough oxygen, and eventually we die.

We can get anemia by bleeding, say from a stomach ulcer or a gunshot wound. We can get anemia because our red blood cells burst prematurely when attacked, for example, by viruses or drugs. Or we can get anemia because we simply don't make enough normal red blood cells because of troubles in our bone marrow.

One of those bone marrow troubles can be in making hemoglobin. Cooley's anemia is a result of that kind of trouble.

Cooley's anemia: too little β globin

Normal hemoglobin holds oxygen the way a baseball pitcher's fingers hold a baseball. Two kinds of hemoglobin fingers (or globin protein chains) called by the Greek letters alpha (α) and beta (β) globin protein chains, and another part of hemoglobin, called heme, acting as a thumb, hold the oxygen in hemoglobin in place, and release the oxygen to the cells in the body as the red blood cells circulate.

Just as an expert pitcher's fingers have been educated over time and through experience to deliver the ball precisely, the structure of the α and β globin chains of human hemoglobin have evolved to interact optimally and precisely with oxygen. But while a pitcher only has to throw 100 or so pitches to the right spots in each game, hemoglobin grasps oxygen from the lungs and delivers it to tissues with perfection, billions of times a day.

Neither α nor β globin chains alone can grip the oxygen. In fact, it takes two α and two β globin chains, together with heme, to hold and release oxygen properly. The normal hemoglobin we have circulating in our blood cells is called adult hemoglobin or hemoglobin A (HbA). Cooley's anemia is caused by a problem in making one key part of HbA, the β globin chains. Without the β globin chains, hemoglobin cannot bind or release oxygen.

The human β globin is a protein like all other proteins, composed of specific chemical building blocks called amino acids. The precise instructions for the amino acid sequence of β globin, again like that of all other proteins, are encoded in our chromosomes by the genes in our DNA. The human β globin gene is located on chromosome 11, at a place called the human β globin gene locus. We carry one β globin gene at this locus on each of the pair of chromosomes that we inherit, one from each parent, so that we have a total of two β genes per individual.

Our two human β globin genes determine the amount of human β globin we produce. Normally, these two genes allow us to produce enough β globin to fill our red cells with normal amounts of normal hemoglobin. However, in Cooley's anemia, little or no β globin is made. A defect in both of our two human β globin genes leads to either too little or no β globin chain production from these genes, and thus to Cooley's anemia. These gene defects in our β DNA are called β thalassemia mutations. If we inherit a β thalassemia gene from both parents, and we have no normal β globin genes, we will end up having little or no HbA. As a result, we will suffer from a severe anemia, Cooley's anemia.

Cooley's anemia is also called homozygous β thalassemia, Mediterranean anemia or thalassemia major. In this book, I will use Cooley's anemia and thalassemia interchangeably to signify the severe form of the disease. This inherited condition affects families in all parts of the world, although it is most prevalent among people of Greek and Italian ancestry who live near the Mediterranean Sea. That is why it is also called thalassemia: "thalasso" is the Greek word for the sea.

If we inherit one β thalassemia gene from one parent, and a normal β globin gene from the other, we have what is called

"β thalassemia trait" or thalassemia minor, a harmless condition with little or no anemia. With the trait, increased production of β globin by the one normal β globin gene compensates for the low or absent activity of the defective β gene.

People with thalassemia trait are not easily diagnosed without special blood tests. They appear normal, and usually don't have anemia or other problems. That is why it is so difficult to know before birth whether a new infant of two parents who each have the trait is at risk for having Cooley's anemia, unless thalassemia trait has been recognized in the parents beforehand, or they already have a child with the disease.

And too much α globin

The low or absent β globin and HbA in Cooley's anemia is bad enough, but there is a lot worse that follows. Despite too little or no β globin, α globin, the other type of globin chain in HbA, continues to be made at its usual normal high level in red cells, and piles up in the cells.

In HbA, the two α and two β globin fingers and the thumb, heme, are all required to hold oxygen in place and release it to the body's cells appropriately. In contrast, free α globin chains alone, without β partners, are useless and harmful. These α globin chains cannot fold properly to grip or hold the oxygen. Instead, the free α globin becomes a tangle of proteins, aggregating as undesirable globs, eventually destroying most of the red cells in which they accumulate.

Our bodies recognize cells with the misshapen α tangles as abnormal, and use scavenger cells to rapidly destroy them. In this process of abnormal red cell production in the bone marrow of Cooley's anemia patients, called "ineffective erythropoiesis," most of the diseased red blood cells, with their α globin tangles in the bone marrow, do not survive.

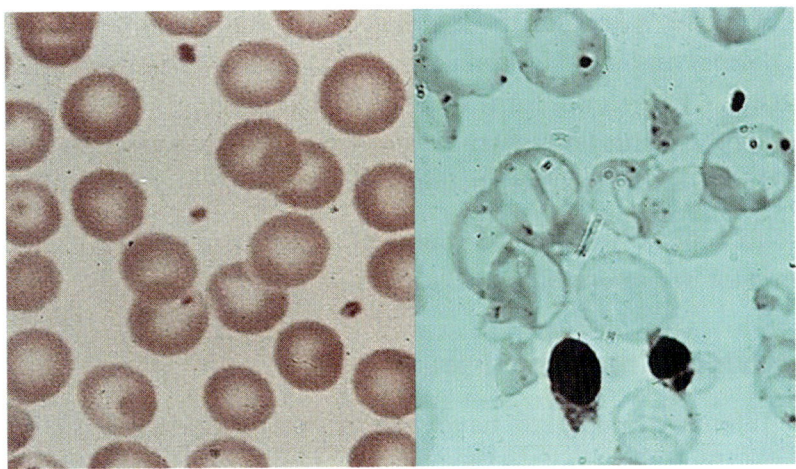

Red blood cells on a blood smear from a normal subject (left), and from a patient with Cooley's anemia (right). The thalassemia red cells are deformed, and contain too little hemoglobin.

Similarly, the circulating red blood cells of Cooley's patients are also destroyed quickly after they leave the bone marrow since they too carry an excess of α globin. The circulating red blood cells in thalassemia appear very abnormal; they are misshapen and have too little hemoglobin.

Too little β globin, too much α globin, thalassemia.

So all one really needs to know about Cooley's anemia is:

(1) The normal hemoglobin molecule is composed of two kinds of protein chains called α and β globin chains, which, together with heme, are necessary to bind and release oxygen.

(2) For hemoglobin to work properly, the two α and two β globin chains have to interact with each other and fold together with heme to form a complete and functional normal hemoglobin molecule, called adult hemoglobin or hemoglobin A (HbA).

(3) In Cooley's anemia, there are too few or no β globin chains produced; so there is not enough HbA. Because of this lack of β globin and the low level of hemoglobin in their blood, the patients have low blood counts or anemia.

(4) In Cooley's anemia, the α globin chains continue to be made in normal amounts, but unattached to β pieces. The free α globin chains aggregate, form tangles, and thus are toxic to the red blood cells in both marrow and circulating blood, resulting in increased red cell destruction and worsening of the anemia.

To repeat: Too little β globin, too much α globin, too little hemoglobin, abnormal red cells, disease.

Surviving to have the disease

Infants destined to have Cooley's anemia survive at birth because they, like all of us, produce another type of oxygen-carrying hemoglobin in fetal life called fetal hemoglobin (HbF). HbF ($\alpha_2\gamma_2$), contains so called gamma (γ) globin protein chains instead of the β globin ones. HbF is the essential and major hemoglobin of all human fetuses. Around birth, the production of HbF is essentially switched off and the production of HbA is switched on. The small amounts of HbF made after birth and into adult life provide only limited oxygen carrying capacity for thalassemia patients.

Why do we need HbF as fetuses? The best explanation for the presence of HbF in fetal life is that it helps the fetus survive, especially when the oxygen supply is limited. HbF binds oxygen more avidly than HbA, like a pitcher holding the ball more tightly. This preferentially allows the fetal red cells with HbF to grab and hold oxygen more tightly than the mother's red cells with HbA, as blood travels in the placenta between mother and fetus.

Infants with Cooley's anemia would die *in utero* if HbA were the only hemoglobin available. The existence of HbF allows their survival.

HbF production is normally turned off around birth in all of us, as β globin and HbA production is turned on, presumably, again, to maintain the fetal oxygen advantage for future progeny. Cooley's anemia results as β globin and HbA production become necessary for oxygen transport after birth as HbF production declines.

We know that patients with Cooley's anemia have normal γ globin genes and normal amounts of HbF *in utero*, and that they survive fetal life normally. So why don't they just revert to making adequate amounts of fetal hemoglobin when they can't make HbA? We don't know why but they can't.

If we could find ways to allow HbF production to persist after birth and into adult life at the high levels it is produced in the fetus, Cooley's anemia patients would be cured. In fact, there are some rare humans who have only fetal hemoglobin in their blood as adults and are healthy; this is a phenomenon that tells us that there is a way to survive without β globin.

The molecular events involved in the switch from human fetal to adult hemoglobin around birth are being better understood and could eventually lead to new therapies. There are already several drugs that are known to increase fetal hemoglobin production. These insights and approaches to regulating and increasing fetal hemoglobin will be discussed later in the book.

Treating the disease

Patients with Cooley's anemia require blood transfusions to combat their anemia in order to survive. When we become anemic from any cause, we have the capacity to spontaneously increase the number of red blood cells that we produce in our bone marrow several-fold to compensate for the anemia.

Similarly, Cooley's anemia patients have this capacity and do increase the number of red cells they make as they try to compensate for their anemia but to no avail. This is because the premature destruction by scavenger cells of most of the defective nucleated red cells with their excess toxic α globin tangles leads to the production of more defective red cells in a vicious cycle that results in a mass of accumulated cells. This mass expands the bone marrow cavity abnormally, and erodes and deforms the bones. The end result of this process in Cooley's anemia is that abnormal red cells fill the bone marrow, but are ineffective in producing useful blood cells with useful hemoglobin.

The abnormal thalassemia red cells that do reach the circulating blood are prematurely destroyed in the spleen, the normal site of circulating red cell destruction. The spleen is the organ that is most sensitive to the presence of abnormal red blood cells. It essentially eats and digests these abnormal red blood cells, engorging and enlarging itself as it does so. The volume of red blood cells consumed increases with the increasing size and capacity of the spleen. This increasing capacity leads in time to the destruction of even transfused normal red cells in these patients. Early removal of the spleen (splenectomy) due to an enlarged spleen (splenomegaly) is common in thalassemia.

In addition, the much more active destruction of thalassemia red cells compared to normal red blood cells leads to an increased release of heme and globin. While the excess globin is broken down into its amino acids and re-used, the excess heme can lead to the deposition in the gall bladder of excess bilirubin which is a breakdown product of heme. This process causes gallstones and gall bladder disease, and may eventually necessitate gall bladder removal.

Transfusions and iron overload

Blood transfusions are required to solve the problem of anemia. New normal blood cells provide new normal hemoglobin capable of normal oxygen binding and delivery. They also suppress the production and the expansion of the bad thalassemia blood cells, and, therefore, can forestall the development of the bone deformities.

If adequate blood transfusions are not available, most patients with Cooley's anemia do not survive past the age of five. Blood transfusions are required for life.

But the life-saving blood transfusions alone do not end the problems of patients with Cooley's anemia. Iron overload and toxicity is another problem.

The metal iron is a component of heme, a part of hemoglobin mentioned previously. It is the thumb in the pitcher's hand that, in addition to the globin fingers, is required to grip and release oxygen from hemoglobin optimally.

Normally, our iron intake is relatively low. As our red cells, the largest source of free iron, break down and release their heme iron, this free iron is either re-utilized to make new hemoglobin or excreted in our stool and urine. These relatively small amounts of free iron are easily bound by special proteins in our blood stream. Whatever free iron there is outside of red cells is kept in check by these regulatory proteins. The deposition of iron in our body's tissues and organs is normally avoided by this iron binding and excretion process.

However, in Cooley's anemia patients, the large amounts of iron in transfused blood overwhelm these normal mechanisms of iron binding and excretion, resulting in much of the excess iron being deposited in almost all organs, most dangerously, the liver and the heart. Sadly, the human body has no good way of ridding itself of this vast excess of free iron. Thus, too many blood transfusions can result in iron toxicity and death. Iron accumulated in the pancreas causes diabetes; in the liver, it leads to fibrosis and cirrhosis; in the heart, it causes scarring which leads to heart arrhythmias, heart damage, heart failure and death.

So patients with Cooley's anemia can die, either early in life from the lack of normal hemoglobin (HbA) in blood, or later in life from too much iron from blood transfusions. Until the 1970s, the treatment of Cooley's anemia was shadowed by the dilemma of a tragic choice: Too little hemoglobin, or too much iron.

A true miracle: controlling the iron

In 1974, a unique new therapy was introduced to rid the body of the excess iron accumulated in Cooley's anemia patients as a result of their blood transfusions. The drug is Desferal (deferoxamine), and it binds to (or chelates) iron. The iron, chelated and bound by

Desferal, is excreted in large quantities in urine and stool and can result in "negative iron balance" i.e., more iron leaving the body than is being taken in.

Desferal is magic: it has dramatically changed and saved the lives of many, many thalassemia patients worldwide. Patients can now be adequately transfused and have enough normal hemoglobin to live normal lives, without having to worry nearly as much about dying from iron toxicity as was the case in the past. This extraordinary drug has both enhanced the lives and increased the lifespan of most patients.

However, Desferal has a major problem of its own: it cannot be taken orally. It can be given intravenously, intramuscularly, or subcutaneously.

Routine intravenous Desferal is impractical, and intramuscular use is painful. Therefore, the drug is routinely given subcutaneously. It has also been learned that to be most effective, Desferal must be given continuously over eight to twelve hours a day, almost every day. This subcutaneous administration, which requires a needle and syringe attached to a mechanical pump, is a delicate and complex procedure.

Continuous infusion is necessary because stored iron appears only slowly in the blood stream over time and Desferal has to be present continuously to eliminate it. In addition, the half-life of Desferal, the time it stays in the blood stream, is relatively short.

For patients like Linda D. who can be compliant with the Desferal program, the treatment is a dream come true and has meant a long life, although at the physical and emotional expense of sticking herself under her skin for day-long infusions almost every day of her life to receive the benefits of this treatment.

But many thalassemia patients are not like Linda D. They have problems with compliance. For these patients, it is just too difficult to adhere to the almost daily regimen of subcutaneous Desferal therapy.

So Desferal along with blood transfusions is the established therapy for Cooley's anemia, but this is a demanding and difficult program. And it is not a cure. Patients on the program are still dying.

More recently, the availability of two iron chelators that are effective when taken orally, named L1 and Exjade, promise to overcome the injections and compliance problems of Desferal. No more pump, no more syringe and needles. Just a pill once or a few times a day. A true blessing, especially for patients who cannot comply with Desferal injections, or do not respond to the drug. The role of these new drugs in overall thalassemia treatment is still unfolding, but their use is already saving lives.

The road to a cure

There should, in fact, already be a cure for Cooley's anemia. We have known so much about this disease for so long that by the 21st century, we should have already found a cure. So I thought 25 years ago, and so I think today. As mentioned, one approach would be to find a way to restore fetal hemoglobin to curatively high levels, thus ensuring effective oxygen delivery, as occurs in our fetal life. More about that later.

There are two other potentially curative approaches to the disease. We know that the only thing wrong in Cooley's anemia is that the blood cells in the patients' bone marrow have a defect in making normal human β globin. And that nucleated red blood cells in our bone marrow normally provide all of the β globin we need. One curative treatment is to transfer the normal marrow blood cells from a person with normal hemoglobin to the patient, so-called bone marrow transplantation; the other approach, human β globin gene therapy, is to provide normal β globin by either correcting the DNA defect in the patient's β globin gene or adding a new normal β globin gene to the patient's own blood-forming bone marrow cells, thus having those cells produce new normal β globin in normal amounts.

Bone marrow transplantation from another person, called allogeneic bone marrow transplantation (ABMT), has already been shown to cure many patients with Cooley's anemia. The new bone marrow replaces the old. Currently, it is the only treatment that can result in a cure. However, the use of ABMT is limited by the need for immunologic compatibility between the marrow blood cells of the donor and the patient. Without this compatibility, a condition called graft versus host disease can result and lead to disability or death. Because of the problem of immunologic incompatibility, there are no suitable donors for most patients with Cooley's anemia, although new *in vitro* fertilization techniques may lead to the availability of more compatible sibling donors. ABMT will be discussed in a later chapter in the book.

The other potentially curative approach, human β globin gene therapy, utilizes either a corrective piece of DNA or a normal β globin gene. In both cases, normal β globin DNA is added to the Cooley's patients' blood-forming marrow cells in a culture dish, and then the cells are put back into the bone marrow of the patient. Optimally, the gene altered cells will produce normal amounts of normal hemoglobin.

Human β globin gene therapy has cured Cooley's anemia in animal models, specifically in mice. Unlike ABMT, gene therapy has no immunological barriers to overcome, and is potentially a cure for all patients with Cooley's anemia. An early stage clinical trial using human β globin gene therapy in patients with Cooley's anemia is currently underway and is discussed later in the book.

*   *   *

This book is about Cooley's anemia: the patients; their families; the research to understand the disease; current treatment options; and the search for a cure. All of the facts are presented as truly and accurately as I know them, and all of the material from interviews with patients, family members and doctors has been largely verified by follow-up inquiries and overlapping accounts. I alone am responsible for any errors that may be in this book.

Part One

Patients and Families

This section tells the true stories of six families with Cooley's anemia during the last half century. All of the patients have the same disease, but the patients' and families' experiences vary widely. Such is the nature of life.

Chapter 2

Love and Loss

"I love you more than one more day." Joan Didion.

"Mommy, am I going to die?" Lisa, age 10, had asked Roseanne, her mother, after seeing a commercial about Cooley's anemia on television. It said early death was the norm seen in patients with the disease that Lisa had. Roseanne couldn't answer any better than she had in the past. "Everybody dies. You remember that little girl of two who was run down by a drunk driver? She died very young. Other people live to be really old, like grandpa and grandma. But all of us die. When we are born, it's not 'if' we're going to die, it's 'when.'"

Lisa was still not satisfied. "Am I going to die?" Roseanne called her husband, Nick, to come upstairs. He said, "We're all going to die," but he was at a loss for more reassuring words.

Roseanne tried again. "When we're born, God doesn't give us certificates with when we're going to die stamped on it. That wouldn't be fair. People would just worry about when they were going to die if they saw the date. So God uses His judgment. Nobody's given a guarantee. If we know when we're going to die, we don't enjoy the living part."

Lisa accepted that as best as she could.

Seven years later, in 1980, on the Tuesday before Lisa died, her blood transfusion had to be stopped at her hospital, New York

Hospital (today, it is part of New York Presbyterian Hospital). She had been weak and frail, "doing poorly for a few months; she had the flu and a high fever," according to Roseanne. In the preceding few weeks, she had persistent left arm and shoulder pain with no diagnosis. The transfusion was stopped because Lisa had "a galloping heart." She did not have a heart attack, but she was in severe heart failure from the iron overload to her heart. She was sent home with a cardiac monitor, and her usual medicines.

Three nights later, on Friday, Roseanne and Nick went out for a brief dinner to celebrate their anniversary. When they returned, Roseanne's sister, who was baby sitting, told Roseanne that Lisa was complaining of increasing "pains all over," in her shoulders, back and arms. She was also increasingly short of breath. Lisa told her mother that she had to go to the hospital. Lisa knew when she had to go to the hospital.

Roseanne called New York Hospital, and they told her to take Lisa to the Emergency Room. Nick and Roseanne put Lisa in the car and drove quickly to the hospital. In the car, Roseanne says, "Lisa was praying. She looked so beautiful."

> Lisa said, as they drove to the hospital, "I don't want you to cry momma. I'm going to die."
> "You can't. I need you, Lisa," Roseanne answered.
> "OK, I'll try to hang on, but only for two more weeks."

At New York Hospital, there were only limited emergency beds available. To Roseanne's frustration, Lisa was admitted at 1 AM to an Urology floor. There were no beds available on the Pediatric floor. Another thalassemia patient whom Lisa knew was also put on the same floor, and he and his father visited Lisa's room late that night. He had broken his arm. Lisa was breathing heavily and unevenly and looked very sick, and the boy stared at her, frightened at her weak condition. Lisa, even in her desperate state, tried to make him feel better.

Lisa asked, "What did you do?"

"I fell and broke my arm," he replied.

"Oh well," Lisa said, "boys will be boys."

Roseanne saw that the children were uncomfortable, and said to the boy's father, "You better get him out of here." And they left.

A nurse took Lisa's temperature in her hospital room that night and said it was normal. Roseanne used "my own test with my palm, and I felt she was burning up." Another nurse came into the room and started combing Lisa's beautiful long hair to comfort her. Lisa told her to stop. She didn't need her hair combed. She hadn't slept for two days, and she needed to sleep.

"Are you her mother?" one of the nurses said to Roseanne, who was in the room. Lisa, in a deep full loud voice that Roseanne had never heard before said, "Why is she saying that to you now? Why is she asking, are you my mother? Of course she is, and a very good mother too." The sound of her voice resonated off the walls.

"It was like the voice from another Lisa, from somewhere else, soulful, from a different world," Roseanne said.

"And you have been a very good daughter too," Roseanne answered. Lisa slept fitfully that night, breathing rapidly and shallowly.

During the night, Lisa became more short of breath. Blood tests were done. Early the next morning, an endocrinologist came into the room and told Roseanne that Lisa's blood sugar level was over 1000. She needed more insulin. Another doctor told Roseanne that, indeed, Lisa had a high fever. He asked the nurse, "How did you take her temperature earlier?" The nurse said, "orally." The doctor reminded her that oral temperatures are unreliable in a patient with rapid breathing. "You should have taken it rectally," he said.

Lisa was immediately brought to the Pediatric ICU. She continued to do poorly. There was little the doctors or nurses could do for Lisa anymore. When she got there, a nurse tried to weigh her. "What are you doing to me?" Lisa asked. "I don't need to be weighed. I'm thirsty." Roseanne's sister ran and got ice chips for Lisa. Nick was there too, trying to comfort her.

Suddenly, with deep sighs, Lisa lifted herself off the bed with arms fluttering like a swan, and then sank into unconsciousness. Roseanne and the rest of the family were told to leave the room. Roseanne said, "Mommy's right outside." She could only think of what Lisa had told her earlier, "I don't want you to cry mommy when I die. You have been a wonderful mother."

Roseanne had answered: "You have been a wonderful child."

After a while, the doctors emerged from the room and told the family that Lisa had died. "When the doctors come out and tell you I died, I don't want you to cry momma," Lisa had said. "She was 17 and died and had never kissed a boy," Roseanne thought.

Roseanne knew that Lisa was very sick over the previous couple of years, and could die. But she did not think her death was imminent when Lisa had gone home from the hospital with the cardiac monitor the Tuesday before. She thought even then that she would be able to help Lisa survive this episode.

She felt that she was misled by the doctors at the visit the previous Tuesday. In addition, Roseanne was reassured in a phone call two days later that although Lisa was very sick, any new medical problems that Lisa developed could wait for Lisa's next visit the following week. "I should have been told then and there that Lisa was that close to dying, and that there was nothing the doctors could do. I would have done anything to prevent it if I knew that, and it was my right to know.

"If I had been told by my doctors that Lisa was going to die so soon, I would have done a lot of other things right then.

"I would have taken her to the dentist to take off her braces so she could run her tongue across her teeth. She loved to do that, but couldn't for the last year because of the steel braces.

"I would have made her the tacos she loved and hadn't had for a while because of her diabetic diet.

"I would have taken her out on the patio in her wheelchair to feel the fresh air every day.

"I had a right to know."

Roseanne felt frustrated and depressed. Her younger daughter had died, and she thought that more could have been done to save her "even if for a minute longer. I treasured every minute with both my daughters."

Roseanne Casamassima was born Roseanne Scopelleti on October 30, 1938. Her mother, also Roseanne, was Irish and fierce. Her father, Thomas, was Italian; his father was from Reggio Calabria, his mother from Syracusa. Roseanne was raised with great love from her parents. She had an older sister and a younger brother.

Roseanne was a beautiful young woman, as I saw from her pictures as we spoke in her apartment in Long Beach, New York this year. She married Nick Casamassima, a New York Transit motor-man, in 1959. His mother was from Bari, Italy, and his father from Naples. Roseanne and Nick always wanted "a houseful of kids running around and being loved."

Soon after her marriage, Roseanne became pregnant with Denise, her first child, who was born on August 26, 1960. Her sister had two healthy children at the time. Denise seemed to be "tired all the time" and would "hit her head with her hand like something was bothering her."

Roseanne had been using her own pediatrician, but Grandma Roseanne was concerned, and pushed her daughter to obtain a second opinion from her own family doctor. "Something is wrong," Grandma Roseanne said. At the visit, when Denise was seven months old, the doctor saw that she was pale and her spleen was enlarged. A blood test showed anemia.

"When she had her blood taken that first time from her finger, it looked pink and watery to me. It didn't look good. A premonition came over me. Something is wrong," Roseanne remembers.

"And I was crying." At seven months, Denise was diagnosed with Cooley's anemia.

Roseanne was told to take Denise to Dr. Marian Erlandson at New York Hospital (NYH). The Children's Blood Foundation, a private entity there, helped support the clinic at NYH, and cared for many children with thalassemia. Dr. Erlandson told Roseanne that Denise had "Cooley's anemia, thalassemia, Mediterranean anemia." Neither she nor Nick had ever heard of the disease, and there was no known history of it in their families. Although she was frightened, Roseanne thought "it's only an anemia, thank God. She'll get blood transfusions and it'll be fixed." She thought it was going to be all right.

It was only when she took Denise to the NY Hospital thalassemia clinic for the first time that she became depressed and anxious. "There were many children there, and some were deformed. They were gray in color. They had prominent teeth. Huge foreheads. Some looked battleship gray. It was horrible to me. And I was 21 years old, and frightened for my baby."

Because veins are small in children and, in those days, needles were not as fine, blood was drawn from Denise's neck for typing and cross-matching. Witnessing this for the first time, Roseanne wanted to cry, but not in front of Denise or the children. It was only when she was alone with the other mothers of patients that she could cry. They told her she could. And she did.

"When I first told my mother what Denise had, she cried out loud, for the world," Roseanne said. "I could never cry in front of my mother, but she cried for both of us." Nick and Roseanne went to a medical library in Manhattan and looked up Cooley's anemia, and they were further depressed by the poor prognosis of patients, "dying around 10 or 11." But Nick and Roseanne thought the material they read was old, and deluded themselves that Cooley's anemia could be managed with adequate blood transfusions.

At that time, she met Frank Ficarra, the founder of the Cooley's Anemia Foundation (CAF), a support group for patients and families, and was invited to his home. He was optimistic about Denise's future, and she was encouraged by their talk.

More important, Denise responded extremely well to her first transfusion. "She giggled out loud for the first time," Roseanne recalls. "She chuckled. We had never heard her laugh before. Her cheeks became pink, and her lips turned red. We had our finger on the trigger. It was going to be all right, we thought at the time."

The ominous long-term outlook for her daughter took hold after a conversation Roseanne had with a doctor at New York Hospital a few months later, when she asked, "When Denise grows up and has a baby, is her baby gonna have this?"

The answer was, "*If* she grows up, the baby may have it…and *if* she gets pregnant, the baby may have it…and *if* she has a baby…"

The portent of that conversation sticks with Roseanne to this day.

"When I heard those 'ifs,' I wanted to hear 'whens,' but I did not. I wanted to pick Denise up and run away with her. To any place where this nightmare might end. I was heart-broken. It was futile. She was going to die too soon. I didn't think it was fair.

"If I myself had been struck with cancer or anything, I could take it. If it was me alone, I could deal with it. But I had brought a baby into the world and she was not going to live a normal life. I always thought it was the most fabulous thing to be a mommy, and to love my babies forever. And I never wanted to stop loving them, and never have. I knew I would love Denise and care for her forever, but I felt so sorry that she was going to die too soon. She wouldn't be ready for that, and neither would I."

Roseanne felt like she was in a cloud that wouldn't lift.

She came home and told Nick the prognosis "was as bad as we read in the library." Her husband, in a rare show of emotion, sobbed and punched a hole in the kitchen wall.

Roseanne weighed about 115 pounds when Denise was born. After finding out her child's diagnosis and struggling through those early months, she had shriveled to 97 pounds. She was comforted by her family's support and love and came out of the doldrums. But instead of resuming a normal life, she became pregnant again. Being distracted by Denise's illness, she had taken no special precautions to avoid it, even knowing that "one in four with every pregnancy" were the odds of having another child with the disease.

She thought about an abortion, but then decided to go through with the pregnancy. "I thought we'd get lucky. The odds were with us. I also made a deal with God. I asked Him not to let me have a boy with Cooley's anemia. In those days, boys had to be tough and strong, and I didn't think a boy could compete in a man's world if he had it. Girls were more dainty then. They weren't expected to compete with the boys. I could protect a girl and love her for her whole life if it happened again. And Denise would have a sister."

She delivered Lisa on May 29, 1962. She weighed only 5½ pounds, but "her color at birth was gorgeous. She was red. I thought 'she's not going to have it.' " She was told at the time that, at most, Lisa would have the trait, that is, inherit only one thalassemia gene from one parent.

But, it is difficult to be sure at birth about the diagnosis. At four months of age, Lisa and her parents went to New York Hospital to be tested more definitively, on a day that Denise was being trans-fused. After the test, Roseanne was told by a nurse, "She's fine. She doesn't have the gene." Roseanne was ecstatic. She called her mother immediately and told her, "Mom, she doesn't have it."

As she returned from the phone call, Nick came up to her anxiously and said, "The doctor wants to see us upstairs." They were surprised and went to see her. The doctor said, "We have to draw more blood on Lisa for testing. There was another child with a name similar to Lisa's, who was also in for testing today, and we have to repeat the tests. I'm sorry. Why don't you go home [it was a Friday afternoon], and I'll call you with the results on Monday."

"No, we'll stay and wait," Roseanne said. As if they could leave. They stayed and waited and the results were that Lisa had Cooley's anemia. "When I heard it I took Lisa who was in my arms, and put her back in her baby carriage. I was angry and felt detached from her for a moment. Then I thought, well, we knew this could happen again."

The family went right to Grandma Roseanne's house. Her mother took one look at Roseanne and broke down and cried. Roseanne's father said, "If I could cut off my arms and take this away from you, I would."

In 1967, the Casamassimas bought a house in Smithtown, Long Island. Denise and Lisa were going to New York Hospital every two or three weeks to be transfused. They were growing normally and had each other. Lisa and Denise had a little game they played while they were being transfused. They received their blood at that time while on hospital gurneys (mobile beds) that were situated parallel to each other in the hall of the hospital. Whoever finished their transfusion first won the game.

One time, a New York Daily News reporter was there to write about the clinic and the patients. In July 1967, to Roseanne's surprise, a prominent article appeared in the Daily News entitled, "A Race For Time," referring to the transfusion game Lisa and Denise played, and obviously much more. The article highlighted New York Hospital, the Cooley's Anemia Foundation, and the Children's Blood Foundation. "It was spread over several pages, and had pictures of Denise and Lisa. They were adorable."

As a result of the article, Roseanne was contacted by other parents of patients, and made a life-long friend, another parent, Maryanne Cervoni. Roseanne also became more active in the CAF from then on, and was subsequently the coordinator for several fashion shows that the CAF sponsored to raise money for Cooley's anemia research.

At that time, 1965–66, I was in the first stages of my scientific career as a hematology fellow doing research on thalassemia at Columbia University. I was searching for the specific biochemical defects in the red cells of Cooley's anemia patients that were

Denise and Lisa Casamassima as children. (All photos courtesy of Roseanne Casamassima).

responsible for the inability of the cells to make normal amounts of normal hemoglobin (HbA). It was unclear at the time what was going on, and these were the early days of molecular biology.

New York Hospital had the biggest and best thalassemia clinic in the city. For this reason, I would go down there once a week to draw blood from the patients in the clinic before their transfusions, and then come back to my lab at Columbia to study how the cells in their blood made hemoglobin.

The children whose blood I used didn't get any extra needle sticks because of my research. I would use the same needle in the arm to draw blood for the typing and cross-matching needed for their transfusions, as well as some extra blood for me to study.

I am sure I drew blood from Denise and Lisa Casamassima. I remember Lisa and Denise's names and that they were sisters. Their blood was used in some very important scientific papers (at least I thought so).

I realize now how relatively unemotional I was about the thalassemia patients and their families back then, even though I had two children about the same ages as Denise and Lisa. I was focused like a laser at the time on my personal goal of scientific success. Though I know I have always tried to be empathetic to patients, I did not allow myself to really feel their plight to the extent that I do today. Maybe back then, by so single-mindedly focusing on scientific and professional goals, I was just doing what most young doctors do: steeling myself emotionally from being overwhelmed by the flood of feelings that accompanies the recognition of the depth and horror of the tragedies we confront dealing with the sick and dying every day.

Childhood for Denise and Lisa was as normal as could be expected given their dependence on blood transfusions every two or three weeks. As far as was possible, their pre-school and school days were happy ones.

As children, they were occasionally singled out in public as being different because of their disease. Children with Cooley's anemia who are inadequately transfused can have facial bone deformities leading to an Asian-like appearance, referred to in medical jargon as "mongoloid facies." Today, these complications are avoided as patients are more vigorously transfused.

Roseanne recalls that once Denise came home from first grade and was very angry that someone had called her "Chinese." Denise told her mother, "I hate you. I want to live with Aunt Marie." (Marie is Roseanne's sister who lived a couple of blocks away.) And off she went to Aunt Marie's, only to return later saying, "Mommy, I love you and I need you."

Roseanne also recalls a time when Lisa, Denise, Roseanne, and Grandma Roseanne were taking a bus to go shopping. An elderly couple seated near the front were staring at the children. Roseanne recalls: "The wife said, 'They're adorable.' Then the husband said, 'yeah, but they're Eurasian.' My mother, trailing us, overheard this and told the starers, 'No, they're Italian. And I'm their grandmother and I'm Irish.' "

Sometimes, the children would smile and confuse the starers by pretending to be foreigners. Roseanne used to reassure them about their looks: "What do you see in the mirror? Look. You. You're beautiful," and hug them, and they knew they were indeed beautiful and loved.

According to Roseanne, once, at a CAF fashion show she organized, a red-headed freckle faced man, at the same hotel for another event, saw Denise, who was then 14 years old. This stranger insisted on telling his friend in a loud voice, "Hey, look at the Filipino." The phrase was repeated several times until Nick finally appeared and calmly told the man that no, Denise was Italian with Cooley's anemia, the disease that the fashion show was about. The man said, "I'm Italian; she's not Italian." Nick said, "Look, I'm her father, I should know," and he forcefully stuck a circular describing the show and the fund-raiser in the man's shirt pocket, and walked away.

Both children were transfused regularly from early in life. When Desferal, the effective iron chelator used to treat iron overload due to the blood transfusions, became available in 1974, Denise was 14 and Lisa 12. They were among the first to receive the drug. They went to the National Institutes of Health (NIH) for three weeks every three months to be studied experimentally on intramuscular Desferal. They were also on metabolic diets and vitamin C to help mobilize iron. The vitamin C therapy eventually was stopped because some patients developed iron toxicity from its use.

They finally received subcutaneous Desferal, the most effective therapy, at NY Hospital when it became available in the late 1970s. Denise and Lisa were fairly compliant with the Desferal, according to Roseanne. "They did it four nights a week on average, not five. And they never stopped completely." I believe they began Desferal too late in their lives to save them from the cardiac toxicity of iron excess.

Diabetes occurs in thalassemia patients due to iron deposition and its toxic effects on the pancreas. Roseanne thought it was the harbinger of death, at least for her children. Lisa got diabetes in 1979, and died in 1980. Denise died in 1986, less than a year after diabetes surfaced in her.

Life up until 1974 had many positives for Lisa. She was linked to her sister by their common disease, and they were very close emotionally. They loved each other; they played together, got transfused together, did most everything together. They both had splenectomies. They both had their gall bladders removed. They supported each other, and their mother, father and grandparents were all there for them all of the time.

Denise was a very smart and observant child. Dr. Erlandson told Roseanne early on, "She notices everything, and knows everything that's going on. Never lie to her. Always tell her the truth." Denise did well in school, was very outgoing, and had lots of friends. Every two or three weeks, of course, she missed a day of school to be typed and cross-matched and transfused. She was an excellent student, and wrote beautiful stories and plays in high school, which Roseanne shared with me.

Denise graduated from Smithtown High, and then started Dowling College in Oakdale, Long Island. But when Lisa died while she was in her second year, Denise decided to "get on with my life," by transferring to Katherine Gibbs, a secretarial school from which she graduated with honors.

Denise, herself, however, also had increasing heart failure and had an acute cardiac episode when she was 21 years old. She was taken to the St. John's Hospital Emergency Room in Smithtown. Denise quite suddenly had chest pains that caused her and Roseanne to be very worried. Denise turned out to have pericarditis, an inflammation of the lining surrounding the heart. Pericarditis is not an unusual complication of heart disease in patients with Cooley's anemia. Patients like Denise sense when they need acute care, and this was one of those times.

The nurse put Denise in a room. Denise was short of breath and very anxious. She wanted Roseanne to stay with her. "You can't be with her. If she can't be in the room alone to wait for the doctor, she can wait outside with you," the nurse said. Denise said, "I need my mother." Roseanne insisted on staying in the room with Denise, and said, "If I have to step over your body to do it, I'll be in the room." She then saw a doctor who knew Denise and knew she had Cooley's anemia; the doctor told the nurse, "Find them a room to stay in together right now."

Three cheers for Roseanne. It takes enormous emotional strength to confront medical personnel when you think they are acting against the best interests of the patient, even for doctors like me when we are patients. Roseanne never accepted the idea that some nurses and doctors can be downright miserable people as they deal with patients, and she would not tolerate bad attitudes in the care of her daughters.

Insensitivity in a physician or nurse is inexcusable. The best physicians and nurses I know are those who think as little as possible about themselves when they are with patients, and, ideally, only think of their patients. That's the oath. That's the way it's supposed to be. Period.

When she was 20 years old, Denise met Eddie, the love of her life. Roseanne desperately wanted Denise to experience love, although Denise, like Lisa, never had menstrual periods and never developed significant breasts. "They were beautiful girls. They knew hair, make-up and clothes, and I did all I could to make them feel as feminine as possible, but there weren't hormones available in the '60s and '70s to help them mature physically," Roseanne says.

"And I was so sad that Lisa didn't live long enough to kiss a boy, so I really wanted Denise to experience love and be with a man."

Eddie was the man. They were together for over five years, until Denise's death. During their first year of dating and courtship, Denise had told Eddie that she had a sister who had died with Cooley's anemia, but she did not tell Eddie that she had it as well. It was left to Roseanne to do that.

Roseanne told Eddie she wanted to speak to him alone. While Denise was visiting a patient friend of hers, Eddie came and

Denise Casamassima at age 20.

Lisa Casamassima at age 17.

Roseanne told him that Denise had Cooley's anemia like her sister. Eddie looked shocked, as expected. He said: "I love her and I want to marry her. I would never hurt her. But I've got to go now." And he walked out the door without another word. Eddie was clearly very upset, and "he probably didn't want to cry in front of me," Roseanne thinks.

Thankfully, Eddie did love Denise, and they eventually became engaged. They went on a planned trip to Canada soon after Roseanne's revelations to Eddie. Earlier, Roseanne had asked the doctors not to tell Denise that she had diabetes because of its ominous portent. Denise knew that Lisa had died within months after a diagnosis of diabetes.

By the time of the trip to Canada, Denise knew about her diabetes, but it was still mild enough so that she was not spilling sugar into her urine. Roseanne had told Eddie about the diabetes, and, although it was still mild, she had told him before the Canada trip to "keep Denise away from too much sweets." The trip was cut short; at 6 am, two days later, Denise came into the house and said, "I have worse diabetes, I'm spilling sugar." Denise had become so anxious by the positive urine test that she had to come home.

Denise and Eddie were married on August 24, 1985. Roseanne was overjoyed. The happy couple went to California for their honeymoon, although Denise was short of breath and had congestive heart failure even on her wedding day. Soon after arriving in Los Angeles, Denise's legs swelled up dramatically, and she called Roseanne who told her to keep her legs up, and eat plums. She also arranged for Denise to get Lasix, a diuretic. The treatment worked and the honeymoon continued. Shortly after returning, however, Denise developed an abnormal heart rhythm (an arrhythmia) that was so severe she needed a cardiac pacemaker almost immediately.

Before she was married, Denise's success at Katherine Gibbs had led her to a job she loved as an executive secretary to a boss at a big telecommunications company. "Her office was high up in Rockefeller Center, and she loved to look down at the Christmas tree when it was up," according to Roseanne. She had been working 60 hours a week for her boss during a busy period before her marriage, but with her increasing heart ailments, she couldn't do it any more.

Denise looked so happy, lovely and peaceful in the wedding picture I saw in Roseanne's apartment. Ten months later, Denise died of the complications of iron toxicity to her heart. She had had more and more heart failure and arrhythmias, and, as with Lisa, nothing more could be done. One of the doctors had told Roseanne, "That golden heart is growing tired."

The doctors at New York Hospital told Roseanne that Denise and Lisa were two of the most severely affected patients that they had ever seen. It may be because different patients with thalassemia either have or lack genes that reduce the severity of the anemia. It may also be that different patients metabolize iron differently. Denise and Lisa also received Desferal too little and too late in their lives to save them.

In her last days, the family and doctors talked about trying to get a heart transplant for Denise, but these efforts were aborted by the wishes of Denise herself.

Eddie and Denise were supposed to go for a vacation at Gurney's Inn in Montauk, Long Island, the week she died. Days before, Denise became bed and chair bound and was getting worse. Roseanne and she ate out on the patio of their house, something Denise usually didn't like to do, but requested. She wanted Hungarian goulash, and ate it heartily and talked about the trip to Gurney's.

Denise said to Roseanne: "Mommy, I'm not ready to die. I'm only married 10 months."

Roseanne replied, "I'm so sorry you were born with Cooley's anemia.

Denise responded, "Don't be sorry. Without Cooley's anemia, we wouldn't have met so many nice people along the way."

She went to the hospital instead of to Gurney's. Roseanne, Eddie and Denise were told that there was nothing more the doctors could do. Denise was in severe heart failure and had arrhythmias that were not controlled by her pacemaker. Her shapely legs were now massively distended and she was short of breath. "I want to go home," she said. "The food is better there." She went home. Eddie massaged her legs and she died in his arms.

Chapter 3

In the Beginning

Frank J. Ficarra founded the Cooley's Anemia Blood and Research Foundation in 1954 because he had two children with Cooley's anemia and he knew that they and other patients with this dreaded illness needed blood. At that time, the only treatment for the disease was blood transfusions and the problem was to obtain adequate amounts of precious blood to keep them alive. From that time to this day, the Cooley's Anemia Foundation, or, as it is often referred to, CAF, has been the major support group, not only helping co-ordinate blood donations for patients with the disease, but also organizing effective campaigns to provide the best and the latest treatments for Cooley's patients, and funding research and education related to the disease.

Frank's son Robert (everybody calls him Bob), a handsome, extroverted, extremely devoted and engaging man, has followed in his father's footsteps. Bob has been President of the CAF in the past, and is currently a major figure in the Thalassemia International Foundation (TIF), a thalassemia support group that reaches 97 countries including the Middle East, Asia and Africa. TIF is a model for translating the success of education and patient care in the US and Europe into better education and care in less medically developed countries.

Bob's grandfather, Antonio, came to the US by boat at the age of 14 in 1898 from Lipari, an island off Sicily. He had only $10 in his pocket. The family had moved to Lipari from Ficarra, a town with the family name in Sicily that General Patton and the US Army had to cross in WWII. The US invasion is detailed in a book called "The Taking of Ficarra." They had moved to Lipari so that Antonio's father could work in the pumice mines there. Antonio's father was killed in an accident in the mines.

Antonio bought the right to marry Catherine Falazo from a foundling home in Brooklyn for $200. Antonio was 22 then, and Catherine 16. According to Bob, the story is told that "About nine months after the marriage, Catherine complained of abdominal pain, and her neighbors told her she was about to deliver a baby. She was astonished. She had no idea about it. Antonio was called from his job at the NY Telephone Company and was amazed to find his son, Frank, delivered by Catherine's friends. It was a surprise to him too. It is said that the umbilical cord was wrapped around Frank's head."

Frank Ficarra grew up in Brooklyn, went to public schools there, and attended Boy's High School. Frank became a butcher in Brooklyn and had his own shop. In 1933, he married Theresa Filippone, from Jersey City, New Jersey. Her family was from Agrigento, Sicily.

Carol, their first child, born in 1938, was found to be anemic. For doctors and patients, the '30s and '40s were the dark ages regarding information about Cooley's anemia. Cooley had described the disease in 1925, but not much was known about it. Some knowledgeable physicians could make the diagnosis of the trait and the disease from the blood smear and the presence or absence of anemia, but most doctors knew nothing about it. A blood smear with oddly-shaped cells and a significant anemia in an infant usually meant Cooley's anemia; an abnormal blood smear without anemia, thalassemia trait.

Carol was diagnosed with Cooley's anemia during her first year of life and was treated at Prospect Heights Hospital in Brooklyn near the family's two-family brick house at 278 East 39th Street. She was only transfused when her hemoglobin level dropped to five or six. The hemoglobin level, a measure of how much blood we have in our circulation, is measured in grams per 100 ml of blood. A normal hemoglobin level is above 12 for women and above 14 for men.

You can get away with a hemoglobin level of 10 without symptoms, but below that, and at levels of five or six, there are often symptoms due to the lack of oxygen; the symptoms are weakness, poor color and, eventually, shortness of breath and heart failure.

Frank and Theresa's second child was Bob, born in 1943, five years after Carol. He was well. Bob recalls, "My brother, Frank Jr. was born two years after me. My parents figured that since I was well, then their next child would be too." They were wrong. Frank Jr. also had Cooley's anemia. He had, like Carol, inherited a thalassemia gene from each of his parents. Three children, two with Cooley's anemia.

Bob's siblings, Carol and Frank Jr., like all patients with Cooley's anemia, required blood transfusions to survive. To help them, in 1954, Frank, together with Joseph Caltibiano who owned a fish store across the street from Frank's butcher shop, and whose daughter Ida also had the disease, formed the Cooley's Anemia Blood and Research Foundation for Children that became the Cooley's Anemia Foundation, CAF, of today.

According to Bob, at that time, a pint of blood (or a "unit") cost $35, a lot of money then, to be paid every two to three weeks. In 1954, the CAF's first major events were blood drives for patients with Cooley's anemia; the blood was collected by the American Red Cross. Over a thousand units were collected in that year alone. In one of these collections, the employees of the Schaeffer Brewing Company donated 151 units of blood specifically credited to the two affected Ficarra children.

The first meetings of the Foundation were held in an office at 16 John Street, around the corner from Prospect Heights Hospital where the Ficarra children were cared for. Frank was the first President of the CAF. Frank soon left the butcher business and went to work for the CAF full-time. The Foundation bought a white station wagon, and paid Frank $10,000 to run a service. He would pick up and

A blood donation poster from the 1950s (courtesy of CAF).

deliver blood in a refrigerated compartment (a box attached to the cigarette lighter in the car). He also escorted patients to and from their hospital visits for transfusions. Blood was collected in glass bottles then rather than plastic bags.

The car also doubled as an ambulance to take patients to the hospital in an emergency. It carried a cardiac defibrillator in case a patient with thalassemia had a cardiac arrest on Frank's watch.

Bob's two affected siblings, Carol and Frank Jr., and several other Cooley's patients were enrolled in Dr. Irving Wolman's "hypertransfusion" program at the Children's Hospital of Phiadelphia, CHOP, then and now a major center for the care of patients with Cooley's anemia. The approach of the Wolman program was simply to keep the patient's hemoglobin between nine and ten grams, instead of the practice at that time of waiting until the patients developed symptoms of anemia, such as shortness of breath, usually at a level of five to seven.

After starting the program, Carol and Frank Jr. received their blood transfusions at Prospect Heights Hospital. Carol was a beautiful girl, according to her brother Bob. Despite her illness and her need for transfusions, "she had lots of friends, went to school, and did well until she was in her 20s."

At that time, Carol had a life-threatening episode of pericarditis, an inflammation of the lining around her heart. In this condition, fluid accumulation can prevent the heart from pumping properly and can lead to sudden death. According to Bob, Carol's life was saved when "this cardiologist came over and stuck a huge needle right between her ribs and removed a lot of fluid from around her heart."

Carol worked in a bank. She was married in the 1950s and adopted a child. Despite her supportive care, she died in 1967 at the age of 28 due to cardiac complications of iron overload.

In the early 1960s, Bob remembers, his father worked tirelessly for the Cooley's Anemia Foundation. He not only transported blood and patients in his station wagon, he also often traveled to Philadelphia to go to Dr. Wolman's clinic. Frank was also intensively involved in fund-raising and growing and governing the Foundation.

Teresa, Frank's wife, worked all of her adult life as a seamstress and made draperies. She continued to work until she was 72 years old, helping support the family all this time. She stuck with one company, Ducraft, because it gave her time off to care for her children when they were sick or needed transfusions.

Bob Ficarra's world collapsed in the space of two weeks in 1965. Frank Sr. died suddenly of a heart attack and heart failure. His funeral and wake lasted five days and many people outside of New York who loved him for all of his good works attended.

Then, eleven days after his father's death, Frank Jr., his brother, succumbed to heart failure. "He drowned in his own fluid," Bob recalls. His condition was the result of iron overload on his heart and "the stress of his father's death." Bob was devastated. He had been active with his father in fund-raising for the Foundation, but with these tragedies, disillusioned and emotionally depleted, he left the Foundation completely for ten years. Bob rejoined in the 1970s, when he was actively recruited by the CAF after he had made a donation in his father's name. He then became the President of the CAF in 1980, and has continued to work for them as an unpaid volunteer, as have all presidents of the Foundation since the early days of Frank Ficarra.

Bob married Marianne Russo in 1965. She is half-Sicilian, half-Napolitano, a background in which thalassemia genes are also prevalent. He and Marianne were "tested" for thalassemia before their marriage. At that time, they were told that they were "fine, normal. You don't have thalassemia." To the contrary, Bob actually does have thalassemia trait, but only found out about it years later, in 1980.

By then, Bob was President of the CAF, and, together with Nunzio Cazzetta, and with the support of the Board of the CAF, had been instrumental in establishing five free centers in New York State that tested individuals of Italian and Greek descent for the presence of thalassemia. He learned then that the definitive diagnosis of thalassemia trait required a special blood test that measured the amount of a minor hemoglobin in blood, hemoglobin A2, which is elevated in most patients with β thalassemia trait.

Bob himself was finally tested at one of the centers, at Columbia-Presbyterian, and remembers Dr. Sergio Piomelli, the director of the center, telling him, "You're a carrier. Your hemoglobin A2 is high." He is disturbed to this day that at the time his children were born, they might have been at risk for having Cooley's anemia, and that he didn't know that.

This situation is a major problem in the potential prevention of Cooley's anemia. Two people marry. There is no known or revealed family history of Cooley's anemia. The future parents assume all is well genetically. Then suddenly, without warning, a child is born with Cooley's anemia. As it turned out, Marianne was not a carrier, and the Ficarras' two children, now healthy adults, are free of thalassemia.

Bob Ficarra went to St Francis College and then to Fairleigh-Dickinson for a Master's degree in International Marketing, and since 1974, has owned his own business, Manrob Textile Corporation. Along with raising his two sons, he and Marianne informally "adopted" a parentless daughter, Linda De Pasquale, in the 1980s. "She had no parents; we had no daughter." I have already mentioned Linda, but she has her own story that I will tell in greater detail later.

During his tenure as President of the CAF, Bob was instrumental in forming the Thalassemia Action Group (TAG), consisting of teenagers and young adults with Cooley's anemia who were active in fund-raising, increasing public awareness of the disease, and dealing with their own unique problems in care. He was also a founder of TIF, the international thalassemia group mentioned earlier.

Realizing how little most doctors, patients, and their families know about the disease, Bob has always championed the education of physicians as well as patients and their families, and the importance of awareness of family histories, especially for those who trace their families to areas of Cooley's anemia prevalence.

I was the chairman of the Medical Advisory Board of the Cooley's Anemia Foundation in the 1980s as well as the editor of several Cooley's Anemia Symposia. These symposia, held every five to six years, were the highlights of Cooley's anemia activity in academia, regularly updating basic and clinical research in the disease. The symposia brought hundreds of basic science and clinical investigators from around the world together to discuss all aspects of the disease with patients and families.

The proceedings of these symposia were all organized and published by the *New York Academy of Sciences* in special volumes, called the Annals, which hold the detailed history of progress in thalassemia research and clinical care, from the first volume in 1960 to the most recent one in 2005. These symposia were initially jointly funded by the CAF and the National Institutes of Health (NIH). More recently, corporate contributions have also been made to this educational enterprise.

Frank J. Ficarra, founder of the Cooley's Anemia Foundation (courtesy of CAF).

The Annals volumes have been a special joy to me personally as a record of my own contributions as well as those of many others to the understanding and management of the disease. I was so impressed with Bob Ficarra's accomplishments, leadership and devotion to Cooley's anemia that when I had the chance, in 1990, I dedicated one of the volumes of the Cooley's Anemia Symposia that I edited to him.

Chapter 4

Nunzio: One of a Kind

Nunzio Cazzetta's parents migrated to the US separately from the Puglia region of Italy along the Adriatic coast, years before Cooley described the anemia that bears his name. In 1912, when he was 26, Nunzio's father, Raffaele, came to Greenpoint in Brooklyn, and he was 35 years old when he married his wife, Carmela. She was the oldest daughter in a Puglia family who was shipped to the US in 1921 for an arranged marriage because her younger sister wanted to get married, and the custom in Italy was that the older sister had to get married first in order for the younger sister to do so. Nunzio's mother never learned English.

Nunzio was born on January 30, 1935, the youngest child in a family that includes an older brother, now 76, and an older sister, now 82. He went to public schools in Brooklyn, and married a 22-year old woman from Elmhurst, Queens named Rose, in 1958, when he was 23. Nine and a half months later, Nunzio Jr. was born to the proud parents. They had no experience with children, but from his earliest days, the infant seemed unhappy, crying often and eating poorly; he also had a "distended stomach."

Dr. Catinella, the family doctor, did a blood test when the baby was four months of age, found anemia, and sent the Cazzettas to Dr. Arthur Sawitsky, a hematologist at Long Island Jewish Hospital. The Cazzettas were told that Nunzio Jr. had a severe anemia and a large spleen, and he would need blood transfusions; they were also told that he "would not live more than a decade . . . although there is a great deal of research going on and things might happen during the next ten years that promise new and better treatments."

Nunzio recalls woefully that "in the next ten years" has been a recurring phrase promising a cure for Cooley's anemia during the last 40 years, many times in pronouncements from people like me.

Dr. Sawitsky proceeded to do a special blood test (hemoglobin A2) for thalassemia trait on Nunzio and Rose, the results of which were abnormally high, and indicated that both Nunzio and Rose had thalassemia trait. Nunzio Jr. had inherited thalassemia genes from both of his parents.

Nunzio Jr.'s diagnosis and prognosis were a total shock to Rose and Nunzio, but they did the best they could. Transfusions were begun every two weeks and Nunzio Jr.'s condition improved. His spleen grew larger, but he met all of his other physical and mental milestones, attended school and had friends. He did break his arm spontaneously as a child, and that was attributed to the bone fragility resulting from bone-eroding expansion of the marrow that occurs in thalassemia.

Dr. Sawitsky monitored Nunzio Jr. closely and the transfusions continued. I knew Arthur Sawitsky well. He was a superb physician with great intelligence and dedication to his patients. But neither he nor any of us could cure Nunzio Jr. then.

The severity of the illness didn't really resonate with Nunzio until the first New Year's Eve after Nunzio Jr.'s birth when he suddenly became particularly pale and was in distress. The Cazzettas took him to a hospital emergency room, and he had to be admitted to the hospital for increasingly severe anemia. From that day on, the Cazzettas did not look forward to New Year's Eve and knew they were in for "a rough time."

Nunzio Jr. was physically active, and even played in Little League. However, at the age of six, his baseball career ended abruptly. In the final game of his first season, Nunzio Jr. was the runner on first base when the next batter popped the ball up to the

Nunzio Jr. (courtesy of the Cazzettas).

third baseman. Nunzio Jr., like most six year olds, didn't realize that he should have stayed on first base, and, instead, ran towards second and was doubled off the base. His coach, enraged, ran out as Nunzio Jr. was leaving the field and started screaming at him for his errant base-running. He came close to physically shaking the boy, despite his knowledge of Nunzio Jr.'s medical condition.

Nunzio Sr. was in the stands and "lost it." He ran out to defend his son and remembers saying to the coach "You sonofabitch, you know what I have to do just to keep him alive!" Nunzio Jr. did not play in Little League again, and Nunzio Sr. was completely turned off to all competitive sports at the amateur and professional levels from that day on. It was one of the few times that Nunzio was really angry in public, and was a rare emotional release for all of the pent-up frustration and grief inherent in his young son's predicament, and his.

Nunzio Cazzetta is one of the calmest, most reasonable, conscientious and sincere individuals I have ever met. In a conference room at the Cooley's Anemia Foundation's offices in Manhattan, he talks to me slowly and with deep feeling about this episode and the rest of his life. He is a physically compact, self-assured, optimistic man with gray hair, a quick smile and self-deprecating wit who has had a busy and varied life.

His plan for his own professional and personal life was shattered by Nunzio Jr.'s illness. He always loved languages and wanted to go into the Foreign Service and travel the world. But he needed to be ready for Nunzio Jr.'s transfusions and other crises all the time. He was not free enough to pursue his dreams, but he was able to express his love of language by becoming a language teacher.

In the 1950s, there was no medical insurance available to cover the bills for Nunzio Jr.'s transfusions and hospitalizations, so Nunzio, Rose, and their sick child moved out of their own apartment and into Rose's parents' house for four years to cope with their medical bills.

Nunzio earned the Master of Arts degree required to teach language in the public schools. He attended St. John's University part-time for five years because it was close to his home. He had to turn down a scholarship for a similar program that he would have preferred to attend at Queens College because it was too far away. He also did tutoring part-time to earn more money.

Nunzio has had a very successful teaching career over the past 30 years. He has taught French, Spanish, Latin, and Italian, all at the same time. He was the Head of the Language Department in the Smithtown, Long Island public schools. Nunzio feels that his whole life has been "to do what God wants me to do." He has been true to this creed.

Six years after Nunzio Jr. was born, the Cazzettas had fraternal twins: Rosemarie and Ralph. There was no genetic counseling in those days, but Nunzio and Rose were not worried. They had been told by their doctors that their chances of having a Cooley's anemia child was "one in four," and they took that literally. They assumed that they had had their "one," and the next three would be OK. Nunzio had a twinge of fear shortly after Rosemarie's delivery when she seemed to have a distended stomach that reminded him of Nunzio Jr. This, thankfully, turned out to be a false alarm and soon disappeared.

It was not until several months later that the Cazzettas learned that Ralph was the unlucky one. He had inherited a thalassemia gene from each of his parents, and had Cooley's anemia. As a result of the lack of clarity in the "one in four" estimate he was given, Nunzio worked incessantly from that time on as part of the Cooley's Anemia Foundation's mission to clarify what "one in four" meant.

He has continued to work with the CAF to the present day and is one of the mainstays of the Foundation. He is a legend in the organization and the Cooley's anemia community for his fundraising, leadership and devotion. He was instrumental in changing the NIH and CAF brochures to read "...the risk is one in four with each pregnancy."

One day, at age 10, Nunzio Jr. fell unconscious while playing on the front lawn. Rose's parents, who were baby-sitting, became terrified and thought he had died. This episode presumably was the result of a cardiac arrhythmia due to iron interfering with the normal electrical conduction system of the heart.

There were no iron chelators available in routine use during Nunzio Jr.'s life. Subcutaneous Desferal was not available then. He did have three intramuscular injections of Desferal at age 9, but the injections were very painful, and were discontinued.

On January 14, 1971 at 2 AM, Rose heard "a horrific gurgling sound" coming from her son's throat in his bedroom. She immediately told Nunzio, who carried him to the car to take him to the emergency room of the nearby hospital.

It was an icy night and Nunzio, in a total panic, almost killed himself and his child by driving wildly, skidding on lawns and streets to get to the hospital as soon as possible. When he arrived, he could not lift his son from the car without help. He screamed and cried for help. He felt angry and helpless. All resuscitative efforts were unsuccessful. Nunzio Jr. died from congestive heart failure and cardiac arrhythmia, the complications of iron overload affecting his heart as a result of blood transfusions. He was 11 years old.

Nunzio had "lots of repressed anger" after his son's death. He had kept careful records of all of his son's transfusions, hospitalizations, complications, and care. Upon returning home the night of Nunzio Jr.'s death, he tore them all up, flinging them into every room of the house. "What the fuck did I keep all these records for? Were they gonna keep him alive?" Rose cleaned up the mess.

With Nunzio Jr.'s death, and Ralph with the disease, the family started to come unglued. "God will take care of us somehow," Nunzio said, after the death of his first son. "There's gotta be something new that's gonna happen, a new medicine, a cure, a miracle," he thought.

For Rose and Nunzio, their son's death challenged their faith. The Cazzettas were good Catholics, righteous people, a close and caring family that had been severely tested by God and fate. Their lives would continue now only with Ralph and Rosemarie.

Ralph's clinical course was less severe than Nunzio Jr.'s. He needed transfusions every two weeks, but he had the benefit of subcutaneous Desferal, an effective treatment, which he took up to five nights a week by pump from the age of 12. He was compliant with this difficult program most of his life, except for a period between the ages of 19 and 24.

During this period, he told Nunzio "I can't win the battle" against iron overload. However, he renewed his chelation treatment after that time and continued with it until his death. He also benefited from intermittent five day courses of intravenous Desferal. "His urine went from rust-colored [the color of iron] to clear," with some of these intravenous treatments, Nunzio recalls.

When Ralph was 13 years old, Dr. Edward Zaino urged the family to have him evaluated in Seattle, by Dr. Donnall Thomas, a world expert in allogeneic bone marrow transplantation, a potentially curative but largely untested treatment at that time. Initial typing to find a compatible donor for Ralph revealed that Rosemarie, his fraternal twin, was the best immunological match; subsequent testing in Seattle showed that Rose, his mother, was an even better potential donor.

Dr. Thomas told the Cazzettas, at that time, in 1978, that there was a "50 percent chance" that Ralph would die from the procedure. With this big number in mind and after praying for divine guidance in a church the family passed while walking back to their hotel in Seattle, all four family members simultaneously said "No" to transplantation.

When they returned to New York, Dr. Zaino asked Nunzio why Ralph had not been transplanted. "If it was my kid, I would have done it," he said. Nunzio replied: "It's not your kid." Dr. Thalia Papayannopoulou, another expert on thalassemia, later told Nunzio that he had made the right decision. She said, "Ralph can live into his 30s or 40s," and he did.

As a young adult, Ralph became as much of a legend as Nunzio in the thalassemia community. He was an amazingly engaging, sympathetic, intense, and loving young man who was, in turn, loved by all. He went to Nassau Community College and became an emergency room and ICU nurse. He was very active in the CAF, and was one of the founders of the Thalassemia Action Group, (TAG).

The group was empathetic with thalassemia patients throughout the world, and Ralph spent most of his time traveling and talking extensively to encourage patients with thalassemia in other communities in the US and abroad not to abandon hope and to be treated with Desferal. He had a great spirit. But he was doomed by his transfusion iron.

To prevent sudden cardiac death, Ralph had a cardiac pacemaker installed in the last year of his life. One night, sleepless, watching television, living alone, Ralph felt a jolt in his chest and wound up on the floor. The pacemaker had fired because of an abnormal heart rhythm. He called the ER at the hospital where he worked and explained what had happened, and was told to get to the ER as soon as possible. He did, and he survived.

Ralph Cazzetta (courtesy of CAF).

Ralph had another attack of cardiac arrhythmia in April 2002 and, again, almost died. Then, a few days before Ralph and Rosemarie's 39th birthday on May 18, 2004, Ralph, who had been hospitalized since late April of that year, developed severe shortness of breath, "asthma" and pulmonary congestion. His general heart function was clearly deteriorating.

A birthday party was planned at his bedside which was to include Rosemarie, her three children (Nunzio and Rose's grandchildren), Rose and Nunzio: the whole family. The day before his birthday, Ralph had increased trouble getting enough air. A positive pressure mask with oxygen was placed on his face to assist his breathing and he did not sleep well.

On the morning of the birthday, Rosemarie and her children came to visit. Nunzio recalls, "Ralph got out of bed, but collapsed and was helped back into bed. A code red was called. Doctors and nurses were at his bedside to perform the emergency procedures that Ralph himself had called out for them to begin. After all, he was an ER nurse and he knew exactly where all this was going.

"From his room, he was moved to the ICU floor on which he worked. While he was there, Rose asked Ralph, although he was in a coma, to live through this day, his and his twin sister Rosemarie's birthday, for her sake. As a nurse and a patient, Ralph believed that in spite of being in a coma, a patient could hear. And apparently he did. He passed away one day after their 39th birthday. That was his gift to his sister Rosemarie."

Nunzio said Ralph's last words were "I'm just not ready for this." Nunzio and I cried when he told me this. We were old friends and we had sat together discussing patients and problems in Cooley's anemia more than fifty times before, over many years, and neither of us had ever cried during any of those times but we both broke down and cried when he told me Ralph's last words.

With the death of their second son, Nunzio and Rose were again devastated. But their surviving family, especially their three grandchildren, is a constant ongoing joy. Nunzio retired from teaching in 1991 and since then continues "to do God's work. God sends you where he needs you." First, he became a trainer of volunteers working with a women's religious group, Little Sisters of the Assumption, in Spanish Harlem. He worked out of an office in a brownstone on 32nd Street between 2nd and 3rd Avenues.

The day after Ralph died in 2004, he was offered a job as a fund-raiser at the Catholic Diocese in Rockville Center, Long Island. Three weeks later, he was on the job and continues to be active at the Diocese and the CAF to this day.

Chapter 5

Connie Paradiso

Connie Paradiso, nee Rosano, was born in Rapino in the province of Chieti, Italy, along the Adriatic, on March 26, 1922. She came to the US when she was three and a half. Her father, Rocco Rosano, had been here since 1915, when he was 15 years old. His poverty in Italy was such that Rocco had come to the US wearing his sister's shoes. As he had refused to join the US Army during the First World War, his family was not permitted to migrate to the US until 1925.

Education was not a high priority in Rapino; most adults were only educated through the fourth grade. For girls, education was given an even lower priority than for boys. Rachele, Connie's mother, had no schooling.

For 10 years, in order to send money to his family in Rapino, Rocco worked as a gardener, day laborer and limousine driver. He and Rachele, by and large, were separated during those 10 years which were also the years of their courtship and early marriage.

Rachele was 31 years old when she arrived in the US with Connie and her five year-old sister, Yolanda. Two years later, Connie's brother, Lawrence, was born. In 1927, with a steadily grow-ing income working in the construction business, Rocco was able to build a detached one family house in Jamaica, Queens for his family.

Connie and her siblings were all healthy. They went to PS 117 and Jamaica High School. Connie remembers a very happy childhood with lots of music in the home and loving parents. Her mother was very determined that her daughters have the educational opportunities that she never had in Italy. There were no hints of thalassemia in the extended family.

Connie started Hunter College and then transferred to Pace Institute where she took accounting and secretarial courses which led to many good jobs. Rocco's business was relatively unaffected by the Depression so the family, while never really rich, lived well. Connie worked for $8 to $15 a week during the Second World War, and finally settled into a job in a yacht-building company, Sparkman and Stephens, in New York City, at which she stayed for seven years.

Connie Rosano met the love of her life, Eddie Paradiso, when she was 18. He was born in 1921 in Woodhaven, Queens, near Connie's childhood home. Eddie was a glider pilot in the military, and spent four years (1941–1945) flying into combat areas and participating in infantry operations in Europe, including Sicily and, finally, Berlin. Eddie's father was an insurance broker, a job Eddie would also assume when he returned from fighting in World War II.

Connie and Eddie courted after the war, and were married in 1948. Connie became pregnant almost immediately with their first child, Susan. Within a few months of Susan's birth, Connie was again pregnant, this time with her first son, Peter. Connie recalls, "Eddie used to say that all he had to do was put his shoes under the bed, and I would become pregnant."

Susan seemed to grow and mature normally, but when she was a year old her pediatrician, Dr. Tosti, remarked that he "didn't like her color." He pricked her finger and examined her blood in his office. "Her hemoglobin was 8.5; she had anemia, and he knew from looking at her blood smear that she had Cooley's anemia. That was the first time that either Eddie or I ever heard of the disease." Here they were with a one year-old daughter who they were told had "a fatal blood disease," and they had another child on the way. They were stunned.

Dr. Tosti suggested that they not regularly transfuse Susan since she seemed to have a relatively stable hemoglobin level of about 8.5, and no symptoms of severe anemia. She had only one transfusion at that time. On hearing the diagnosis, the Paradisos sought other opinions. They consulted with one doctor who wanted to "give Susan iron pills and blood transfusions every two weeks." Giving iron to patients with thalassemia is always a bad idea since they already have too much iron in their bodies from their blood transfusions. The Paradisos decided to stick with Tosti and his conservative approach.

The Paradisos' lives were changed by Susan's illness. Eddie, who had thought about going to law school, kept selling insurance instead. In addition, Eddie's father had died of a heart attack two years earlier, and Eddie was left to care for his mother as well.

But Eddie did well in the insurance business and, with the help of Rocco, the Paradisos had their own home in Garden City, built in 1953. Rocco and Rachele also built a house for themselves next door. "Peter, thank God, turned out to be well; he only had thalassemia trait," Connie says.

The Paradisos had a solid and devoted group of family and friends who supported them throughout their lives, especially after they learned about Susan's severe illness. Both sides of the Paradiso family, especially Connie's brother and sister, were very helpful, along with a large number of childhood friends. And they had Dr. Tosti, who was "a jewel of a physician."

There were varying views about how to deal with Susan's illness publicly. Rocco thought it should be "kept quiet." It was a black mark on the family. This was not an unusual position to take for families in the '40s and '50s (and even today, to some extent) with children suffering from Cooley's anemia. By contrast, most of Connie's immediate family and most of her friends were more outgoing and favored making Susan's disease known.

Rachele was a feminist far ahead of her time. She had urged Connie and her other daughter to manage all of their pregnancies "in their own best interest." Even before Susan's illness was discovered, Rachele "was livid" when she found out that Connie was pregnant again with Peter right after giving birth to Susan. She thought that women could and should control their own reproductive destiny despite being Catholic, and she urged Connie not to have any more children.

Susan was doing well until age two when she developed a very large spleen that made her feel very uncomfortable. She had a splenectomy at that time and felt much better. The fact that Susan was not getting blood transfusions regularly, which would ordinarily suppress the production of the abnormal thalassemia blood cells, probably accounts for her big spleen so early in life. On the other hand, Dr. Tosti's decision not to transfuse Susan regularly certainly delayed iron overload issues.

A third child, Janice, was born to the Paradisos in 1954 and was "normal." Luckily, like Peter, she only had thalassemia trait, inheriting only one thalassemia gene from either Connie or Eddie, but not from both. By this time, Susan was growing poorly. She was small, quiet, and somewhat frail even though she was not severely anemic. She was now taking folic acid and receiving occasional transfusions.

The reasons for Susan's relatively well-compensated blood production, that occurs in other patients as well, were a mystery. There is a clinical condition called β thalassemia intermedia in which patients are anemic, but less so than in Cooley's anemia. This condition can result from the inheritance of two "mild" β thalassemia genes that lead to enough β globin so that transfusions are unnecessary. However, the fact that the Paradisos later had a child, Paul, with severe Cooley's anemia makes it unlikely that any "mild" thalassemia genes were present in this family, since Susan must have inherited the same β thalassemia genes as Paul did.

Some other genes must have been involved in making Susan's disease less severe. My best guess is that along with inheriting her β

thalassemia genes from both parents, Susan also inherited another gene from one or both of them that decreased the amount of excess α globin and ameliorated the severity of her anemia. This could have been a so-called α thalassemia gene that is known to decrease α globin output. Alternatively, another gene we still don't know about, that affects either red cell metabolism or its life span, could have been responsible for her milder disease.

In 1956, Connie became pregnant again, with Paul. With Susan being frail and obviously ill, Connie was fearful about the result of the new pregnancy. She "did not want to bring a second child into the world to suffer like Susan." She had two healthy children and did not want to roll the dice again. But she didn't even know the odds. One doctor had told her that "you already have one child with Cooley's anemia; this new baby will be fine."

As I've said before, at that time, it was not widely known that in thalassemia, as in all genetic diseases in which severe illness is only expressed when two copies of the bad gene are inherited by the patients, every pregnancy carries the same risk: one in four, of having an affected child. This holds true no matter how many previous children you have with the disease. Every pregnancy! It's like flipping a coin. Even though you may flip a coin 99 times and get 99 heads, you still have that same old one in two chance of it being heads on the hundredth flip. It's the law of chance as well as of thalassemia genetics.

Connie was so concerned during her pregnancy with Paul in 1956 that she visited a clinic in Brooklyn which provided abortions for medical reasons. At that time, clinics required three physicians to certify that the procedure "was medically indicated and that I was competent to make this decision." Connie never followed through.

With Paul, Connie's worst fears were realized. Before his first birthday, Paul was diagnosed with Cooley's anemia and required regular blood transfusions. He had inherited the same β thalassemia genes as his sister, Susan, but he had much more severe manifestations of the disease. Paul, unfortunately, presumably lacked the inheritance of other genes that reduced the severity of Susan's anemia.

With Paul's diagnosis and his transfusion requirements, the Paradisos went to their first Cooley's Anemia Foundation meeting that year. "It was extremely depressing to us," Connie says. Frank Ficarra was the founder of the group. Two of his children and a few others in the group with the disease were older. Some of them had severe facial bone deformities, dark skin from transfusions, and brittle bones, the results of marrow expansion and relatively few blood transfusions suppressing the bad red cells.

All Connie could think of was, "Is this what I have to look forward to with Susan and Paul?" She was depressed. But the CAF was also a place for camaraderie and support, and it became a positive focus of the Paradisos' lives as well as an important and distinct morale booster from then on.

The Foundation's meetings were times when Cooley's families could share information about treatment and research, and, most important, experience that camaraderie and mutual support. Together, they were much stronger than as individuals. They could begin to lobby for patients with Cooley's anemia to receive federal and state aid: for improved blood transfusions; for education of affected families about optimal therapy; for increased awareness of the disease and its complications; and for support of new research. These were all things that were new to the Paradisos. They became instrumental in spreading the word throughout the wider Cooley's community back then, and, in Connie's case, to this day.

The CAF has grown and endured as the dominant force in combating the disease over this past half century. In 1960, the Paradisos established the Long Island chapter, the first chapter of the Cooley's Anemia Foundation outside of its central headquarters in Manhattan. This chapter had some of its own financing and activities, all the while also participating in the "central office" CAF activities. It served as a model for the development of other CAF chapters throughout the New York-New Jersey region as well as nationally.

Despite the severity of his illness, Paul was very outgoing. He loved sports; he gave talks about his disease; and he sold raffles for the Foundation. He was the best man and gave out cigars at Peter's wedding in 1973.

Then that fall, Paul suddenly needed hospitalization for heart failure, and was subsequently in and out of the hospital. A family friend had regularly given Paul tickets to baseball games, and that year the Mets were in the World Series. Paul came out of the hospital and, with effort, he went to a game, and he enjoyed it enormously. Shortly thereafter, sadly, he was gone. After returning to the hospital, he died peacefully there. Paul succumbed to iron overload involving his heart at the age of 17.

Desferal, the first effective iron-chelator which has prolonged so many thalassemia patients' lives, became available for use in 1974. Too late for Paul.

Susan continued to do fairly well with few transfusions until, as a teenager, her brittle bones and poor growth and development led the Paradisos to try the new regimen of "hypertransfusion" initiated by Dr. Wollman at CHOP in Philadelphia. Susan began these more regular transfusions at age 13, and responded remarkably well. Connie remembers: "She suddenly grew into a full woman. Her bosom development particularly was the talk of everyone." Susan finished high school, went to Nassau Community College, and then obtained an Associates degree in Medical Technology and a Bachelor of Sciences degree at Adelphi. She also taught developmentally delayed children in her free time. She began Desferal when it became available, in 1974.

Susan became severely ill at age 26. She developed gallstones, a complication of any anemia with excess blood destruction, including Cooley's anemia. This complication is due to the accumulation of bilirubin, a by-product of heme; bilirubin deposited in the gall bladder forms the stones. Because of her pain and stones, she had her gall bladder removed.

Connie feels that Susan never fully recovered from the surgery and its complications. Her abdomen became bloated and full of fluid (a sign of liver failure), and she required a tube to prevent the fluid from accumulating again. Then the effects of her iron overload appeared to overwhelm her heart as well. She died in 1978 at the age of 28.

"Unlike Paul, Susan suffered for two years in and out of the hospital, after regaining a full life and after much improvement of her condition with transfusions and Desferal," Connie says. "There's nothing as bad as losing a child, but in Susan's case it was worse because we thought she was doing so well."

Susan had helped Connie weather the recent tragedy of Paul's death. When Susan died, Connie had lost two children in five years, and she no longer had the good friend that Susan had become to her. Connie often went to church with Susan. After Susan died, Connie took a "time-out in church," one hour weekly to think about Paul and Susan.

Dr. Tosti had been like the Paradiso family's patron saint. He made house calls throughout the children's early life. Connie remembers him coming promptly when three of the children were sick with various ailments. He charged almost nothing for the care of the Paradiso children, and he always offered good advice.

Sadly, he had moved to Schenectady by the time Susan became ill with her gall bladder disease. When he came to Susan's wake, he told Connie "Susan didn't have to die. Something bad happened with the gall bladder operation." Hearing this from Connie, I didn't think it was such a good idea for him to tell her something like that. However, Dr. Tosti's support and medical care for the Paradiso family far outweighed the unnecessary heartache he caused Connie with that remark.

Formal religion has not been a great part of Connie's life, but she clearly has religious faith. This faith is placed in people, in family and friends. Her relationships here on earth are what comfort her most, and allow her to survive and be happy.

Being Catholic and unrepentant regarding her feelings about terminating unwanted pregnancies, Connie did not feel that she could go to confession "with a clear conscience." But she remembers at least two priests who understood her dilemma about bringing children into the world who might be stricken with Cooley's anemia. One priest, early on in her married life, told a friend of Connie's, "I can't blame her. Until you're in her shoes, you can't judge her."

More poignantly, Connie still remembers at a later time being told directly by another Catholic priest that, "If you are sorry you offended God, that is enough of a confession to make, and to receive communion."

After Susan's death, Connie thought "life would calm down" for her and her family. Peter and Janice were beginning successful healthy adult lives, and Eddie was doing well in business. Their activities in the Cooley's Anemia Foundation were flourishing and rewarding. Then, her husband Eddie's sudden death in 1985 was another major emotional blow. As he had predicted ruefully to Connie sometime earlier, Eddie died of a heart attack at the same age, 63, as his father did. Now she was alone.

Connie had developed "bleeding ulcers" in 1974 and over the next 20 years, she was plagued by multiple recurrences. Finally, in 1993, she had surgery to treat the ulcers.

Connie is 85 now, and, as always, strong, with a full round smiling face and big green eyes. She is still a remarkably lucid and active woman. She lives alone in a lovely condominium in Garden City, Long Island. Her apartment is filled with pictures of her family. She showed me two wonderful ones of Paul and Susan as youngsters. They were fine-looking children.

Her two living children have given her great joy, both in themselves, and in the five healthy grandchildren that they have provided. Her son Peter, a PhD., is a biochemist who works for a major pharmaceutical company, and has three sons. Janice, her daughter, runs a physical therapy clinic with her husband, and has a boy and a girl. The whole family has worked hard for the CAF for many years.

I must have given talks to the CAF at least 30 times in my career as a scientist and as the head of the Medical Advisory Board, and I always saw Connie and her children there. They were grateful for what the doctors and scientists could provide. What we could do was never enough, but we did the best we could.

Today, Connie is gratified by the benefits of medical progress. She especially likes the availability of the new oral iron-chelators that, like Desferal, promise to prolong the lives of people with thalassemia. She has fostered basic and clinical research activities all her life. She is proud that basic and clinical research in thalassemia is where most of the money raised by the CAF goes, although the Foundation is extensively supportive of patient care as well.

Now she is concerned that the CAF is getting a little more impersonal. She says, "It used to be that the Board met once a month all together and everyone was there. You remember that when you were head of the Medical Advisory Board, don't you? Now, there are more teleconferences and videoconferences. It's less personal, less supportive and emotional. But it's still the best there is for Cooley's families."

Connie Paradiso.

Chapter 6

A Life on the Run

Linda De Pasquale is a survivor by all criteria. She has survived Cooley's anemia amazingly well for 45 years, one of its longest survivors. Even apart from the burden of having Cooley's anemia, her life has been one of great emotional turmoil. Yet, as she looks out the window of her condominium near Sandy Hook, New Jersey, and sees the endless Atlantic in its majesty on this sunny day, she is mostly smiling and upbeat. She is grateful for her life so far, and determined most of all to continue to enjoy it on her own terms.

She has been receiving blood transfusions every two weeks for the past 44 years. And she has been injecting herself for 30 years, since 1977, with Desferal, the iron-chelating drug that has kept her iron levels within reasonable limits and thus kept her alive.

"I first started getting Desferal intravenously for a week on and then a week off for a month, and then subcutaneously." She is mindful of all who have since passed away. "Twenty years ago, there were several of us together getting transfused at New York Hospital. Today, there are very few older patients left. Why is that? I don't know."

I tell her, "Neither do I."

"There must be different forms of Cooley's anemia and maybe I have a mild one," she says. I tell her that I don't think so. "I have no sure answer but anyone who has required as many transfusions as you, and had a sister with the disease, must have the severest form," I say. And she and I remain dumbfounded as to the why of her longevity, but both of us are happy and smiling that she is alive and well as we look out, from her new apartment, at the blue Atlantic, past the sandbar that is Sandy Hook. We see Long Island off in the distance, almost like a different world far away.

Linda has migrated from her two-family home in Bellerose, Long Island to finally land near Sandy Hook, with many short stops elsewhere along the way.

"My disease has made me strong and courageous. I can't stand doing anything that doesn't make me happy for very long. I don't have the time for it. When you think your life is going to be cut short, you don't want to waste time on things that don't make you happy."

Time has miraculously been on her side. Her hemoglobin is well controlled by transfusions, and the iron levels in her body are kept low by Desferal. "I don't know for sure about my heart, but I can only hope it's doing OK. What can I do, anyway; it's either going to continue to be all right or I'll die. We're all going to die sometime, but I'm not going to worry about it.

"Somebody said, I think it was either Dale Carnegie or Norman Vincent Peale, 'Worry is a wasted emotion'." Her cell phone rings. It is someone from her large circle of friends and family who fill her life every day. She says she'll call back later.

Linda has worked most of her life, mainly in finance with Chase Bank and as a broker with the bank. She was rolling along in the business until 2001 when she "ran out of gas. It was just too much. I was working five days a week in lower Manhattan. Every other Tuesday, I had my blood typed and crossed at lunch time, and on Thursday, I got my two units of blood after work, and still had to function like everyone else the next day.

"It got to be too much between work demands and health demands. I went into the hospital in early 2001 for a study of a new drug on a new protocol. I felt burned out. Then when I got sick in July 2001, I had an epiphany: It was time to retire. I had worked hard enough and long enough and I was worn out completely. I thought I deserved to enjoy a retirement.

"So I finally decided to go on permanent disability. If I didn't have thalassemia, I probably could have moved up the corporate ladder in the financial services arena, but that's about my only regret. Thalassemia has always been part of my life, and it's given me lots of friends I would never have otherwise. And lots of unique experiences too."

Now Linda is taking college French 2; she wants to go to Paris soon. And she spends a lot of her time with friends and family. She is continuing her transfusions and her Desferal, and is resting more. She also writes very well, and still has the energy and spirit to be optimistic about her life.

Stricken genetically with Cooley's anemia, a terrible disease, Linda was blessed with a wonderful and caring mother who always told her that she was "the best thing that ever happened to her." Her mother eased Linda's path to medical care, emotional health, and growth and strength of character.

Her mother was there for her until the age of 21. And since her mother's death, Linda has continued to do whatever she needs to do to survive, to this day. She is a compliant and compulsive patient who does "everything I can to be healthy. I have all my life. I would never abuse my body in any way. I eat carefully and I keep up with my transfusions and Desferal. I do whatever I can and have to do to stay well."

Linda had an older sister, Nina, who died three years before Linda was born. As a child, she was told by her mother that Nina had died of "heart disease," a condition that her father had died of when Linda was eight months old. It was only at the age of 19, and through a tenant's accidental disclosure, that Linda found out that Nina also had Cooley's anemia. "In the '50s, there wasn't much anyone could do for the disease." Her mother had protected Linda from the facts of Nina's illness so that Linda would not be pessimistic about her future life and her potential fate.

Linda says, "My mother was both my mother and my father. She was my ultimate gift in life. The foundation for the rest of my life. And when she died, my aunt was another wonderful woman, a second mother to me." Linda has a sister, eight years older than her. She was a friend during her childhood, and cared for Linda while her mother worked as a waitress.

The two-family house her mother owned was always the family's prize possession. It was also Linda's home throughout her youth. When her mother died in 1983, she left the house to Linda and her sister. Her sister married one month after her mother's death. The marriage forced Linda to move out of her own home and into her aunt's house for four months because a tenant, whose lease had not expired, occupied the upper floor of the house at the time.

Linda eventually moved back into her family home. But life there became turbulent from that time on. Linda lived on the upper floor. Her sister, together with her husband and their two daughters, and later, after a marital separation, her sister's boyfriend, lived on the lower floor. In 1991, Linda left the house, although she was a half-owner.

She moved to Las Vegas. "I needed a change," Linda says now. She loved the life in Las Vegas. She worked as a change girl at the Rio Hotel and Casino and was on the graveyard shift for a period of time, until she could no longer take the physical demands of that life style. It was a hard life for a young woman who was being transfused every two weeks and taking her Desferal subcutaneously by pump at least five days a week.

She stayed in Las Vegas for a while longer at her cashier's job after being transferred to the day shift. She also started a course working towards a master's degree in communications at UNLV. She thought that the rent payments from her part of the Bellerose family house would provide enough financial support without her having to work full-time.

But she and her sister then decided to sell the Bellerose home. Linda had to return to New Jersey. The house was sold and Linda did get some money from the sale. "My pattern has been, I either run out of money or I run out of energy." In fact, though, she has done extraordinarily well in life, considering the severity of her illness. She is still extremely close to her sister's two children, her two nieces, who often come to visit her.

Linda was very happy as a child, and in grammar school she had lots of friends even though she was underdeveloped, small, sickly, and pale. "I was a happy-go-lucky kid." But when she was transferred to a junior high school in another town, she was not treated well by her peers. "Once they dunked my head in a water fountain. That was the only bad physical thing that happened to me. But they used to say very insulting things to me and it made me depressed."

At that time, she had a protuberant belly due to a big spleen, and they used to say, "Are you pregnant?" But she always bounced back emotionally, despite her considerable physical disability. She was tough and resilient. After high school, she went to St. John's from 1980–1984, and attained a BS degree in business studies.

When she was 14 years old (the age at which her sister Nina had died), Linda's mother took her to Lourdes, the Christian pilgrimage site, in Giverny, France. At that time, Linda was scrawny and underdeveloped, "without my period or boobs." While not particularly religious, her mother had taken her to Lourdes to pray for some miraculous intervention that would positively transform Linda's life. "At Lourdes, I was so small and thin, they thought I was a boy and put me on the boys' line."

When Linda was alone with a nun in the traditional solitude of the holy baths at the site, she stepped into the "freezing water" and was told to make her request (to God and St. Bernadette). "Instead of praying for a cure for thalassemia, I prayed for boobs. I asked for a 36C." Her mother was quite upset and told her she was crazy when Linda told her this. "I didn't lie to my mother very much." Eventually, they both laughed at the experience.

As it turned out, in the next two years, as if by a miracle, Linda developed normal secondary sex characteristics: breasts and periods. While this might have been due to divine intervention, it was more likely the result of her recent start of a "hypertransfusion" program at about the same time which kept her hemoglobin at the nine to ten gram level (much higher than the usual six or seven). She also began Desferal about this time in 1977–1978. It was all like magic to Linda, whatever was responsible for her new femininity.

Religion has never been an important part of Linda's life. She was baptized and made her communion when she was young, but she never went through the confirmation process at the customary early age because when she took religious instruction her peers "made fun of me and the teacher ignored it." She was finally confirmed in 1996 when she was living in Hoboken, during a brief period of trying to connect with Catholicism.

Two months after that she met her future husband. She was married in 1997 to "the love of my life." She thought she had found her life soul-mate. However, the marriage was relatively short-lived because "I found out things about his lifestyle after we were married that were unacceptable to me."

Linda has been a patient at New York Hospital for many years. Among her doctors were Dr. Virginia Canale; Dr. Elise Markenson; and, for the last 30 years, Dr. Patricia Giardina. There was a social environment at New York Hospital. Several patients had lunch together and had their transfusions together; and their families knew each other well.

Many in Linda's early circle of Cooley's friends have now died. She is still a patient at New York Hospital, and considers the hospital and the doctors there very important in her life and part of the reason for her long survival. In recent years, she has been surrounded by a younger generation of patients who have been "very uplifting" to her in their positive attitudes to the disease and to life.

She has had many positive experiences through all her medical travails. She knows "It'll all work out in the end or I'll die. If you're dead, you have nothing to worry about anyway." And, in any event, in the end, she doesn't think "this life is 'it'." She believes in the philosophy of Buddhist reincarnation. She had a dream recently that did, however, frighten her. A dream that she had cancer. She felt "it was real life. Now it was right in my face. And I was afraid. At some time or other, we're all faced with that."

Happily, Linda has not been faced with many life-threatening situations to date, in real life. She remembers once, though, as a young adult, when she thought "it might be the end," going into New York Hospital with a high fever and severe weakness. "A blonde nurse took my temperature and told me I had no fever. I thought she was crazy. I knew my temperature was 105! I was startled. I got paranoid.

"I signed out against medical advice as soon as I could. I thought they were going to kill me. My aunt took me to her house and I recovered. Who knows what it was? I just didn't think the hospital was doing the right things. It was like a comedy scene at the end, with me walking out of the hospital followed by my boyfriend followed by a doctor screaming at my aunt, 'She's gonna die if she doesn't stay.' I didn't die. I got better.

"I am very paranoid when it comes to my own care. I believe in the transfusions and the Desferal, but I don't really like to try new things, since my ferritin, my iron level, has been at an acceptable level for all these years with this treatment, and I seem to be doing OK."

I ask why she isn't taking the new oral-iron chelating drugs, L1 or Exjade. Did she know about the potential value of taking Desferal and L1 together? She knew that, but says, "I'm doing OK with just Desferal." She has been on many drug trials at New York Hospital, and recently took Exjade for 19 days.

"I didn't want to take Exjade, even though it's oral, because I got a little rash on my leg as a side effect and I'm happy with what I'm doing. I also don't like it when there's big money behind new treatments." The pharmaceutical giant, Novartis, produces Exjade which has FDA approval. There have been recent reports of kidney problems with its use.

Linda has a long-term perspective on potential cures. She wants them to come, but thinks "it will probably take decades for that to happen." She told me gene therapy "will probably work at some time, but I'm not waiting for it to happen." She's happy with her current treatment, although she is well aware that time is at a premium for her.

"This disease makes you brave and courageous. If I can stick a needle in my stomach every day and use the pump all night, I can do anything. I've never really had much control over my life, but I've been really lucky. Look at my luck with blood transfusions. There were always anonymous donors. Early on, who knows who they were?

"There was hepatitis. Then, there was AIDS. It was like Russian roulette getting blood in the '80s before testing. When I can control something in my life I do, but when I can't, I don't worry about it."

She is "not much of a joiner," but had a close relationship with the Cooley's Foundation from 1985–1995. After her mother died, she had written a letter to Bob Ficarra, then President of the Foundation, saying she "wanted to do something" for the Foundation.

He never saw her letter. She then happened to sit next to him on a plane to a TIF meeting in Cyprus. "I never knew my father, and my mother always told me that when I was a child, she would always find me sitting on some older man's lap. I was looking for a father and I found one in Bob."

The Ficarras, especially Bob's wife, Marianne, were always there for her when she needed them. When she returned from Las Vegas in 1993, her finances and her life were in disarray as she tried to negotiate the sale of her home with her sister. She moved in with the Ficarras until her life straightened out the next year. The Ficarras treated her like a daughter.

She called upon the Ficarras in a medical emergency in 2001 when she felt poorly and had a blood test done at St. Mary's Hospital in Hoboken that showed "a sky-high white blood count." She was so panicked by the result that she had the Ficarras pick her up by car in the middle of the night and deliver her to the New York Hospital Emergency Room where they stayed until she was admitted. She was diagnosed with severe pharyngitis.

There was black comedy associated with that visit to New York Hospital. Katie, Linda's cat, was included in the hospital trip. Linda, despite her own malaise and a high fever, was primarily concerned with being reassured that Bob was caring for her cat as she waited for admission to the hospital. Bob did his best while chasing the cat around the bushes on East 68th Street.

It was during that July 4th weekend that she spent in the hospital in 2001 that she had her epiphany. She decided it was "time to retire. I had worked hard enough. I was tired." And since then, she has been enjoying her Atlantic Ocean view, her friends, and a life much less on the run.

Linda is a talented writer.
Here is an essay reprinted with her permission.

Can't Purge the Wizard by Linda De Pasquale

I am supposed to be on the 1:15 pm bus to Atlantic City with all the old fogies, but I decided against it, in most part due to this essay. This probably was the day I would have hit a royal flush, and I would have been able to purchase my much needed slider for the terrace. That's ok-I believe in destiny, and it will happen one day. Spoken like a true gambler.

It's been a while since I have been there, at least three months, so I am due. I promise an essay covering the bus adventures and the odd species of mankind that are Atlantic City bus regulars.

Goiter man will definitely be covered, along with the wealthy ladies who like to slum it for the day. I make believe I am a journalist on assignment. It makes it feel better in my head. Anyway, back to the subject at hand.

While I was changing Dean's litter, which happens to be in the laundry room, for lack of a better place, well hidden from view, I decided to clean up the whole room, which is approximately eight feet by six feet, so you wouldn't think this would be a big deal. In addition to it being a small area, it houses a washer, dryer and hot water heater, so there isn't a lot of available cleaning space. This room also happens to house the "Oh my God, what do I do with this?" items which are presents from well-meaning friends and relatives. It is the Room of the Nightmare Chachkas. What a good title for a B horror movie.

I actually have a beautiful shelf holding some of these chachkas, along with wall space being taken up with chachka pictures, and the top of the hot water heater occupied by various and sundry and insane items. I have to say some of these horrors were actually purchased by me in a weak moment. Like the beanie baby cat that used to sit on top of Katie's booda litter box. I survey these trinkets at least once a year thinking that they must be purged.

The Wizard was almost purged today. He's a pinkish red candle in the shape of a wizard and he holds a heart in one hand, and flowers in the other, and he looks happy to see you. I have had the Wizard now since September of 1993. He has survived three moves. He is the ultimate dust collector. He has been efficiently collecting dust now for almost 14 years. Yet, I can't do it. I can't purge him. So unlike me, since living here I have honed my heartless purging skills to an all time high. Still, I can't part with him. He is symbolic of the love that my friend Melissa has for me. That is the most important part. She gave it to me to remember her when I left Las Vegas after living there for two years and having the pleasure of her friendship. I just can't purge a chachka with such strong ties to the heart.

The Wizard isn't the only chachka I can't purge. There's the good old pink, yes pink, 8 × 10 picture of Our Lady of Guadeloupe. This definitely would be a highly esteemed piece in the Mexican barrio. When it was given to me, it literally took my breath away, in the "oh crap, not again" way.

But I ask you this: How do you throw out the Virgin Mary? You don't. So she is on the wall staring at my hot water heater, praying for its longevity and salvation. I believe in the miracles the Virgin Mary can do. I experienced it first hand from the Lourdes "pray for big boobs and you'll get them" experience.

That brings me to the January angel. Yes, a trinket from a frugal friend, or somewhat of a friend, that probably cost about $1 in the dollar store. She is an angel after all. With halo and all. The clincher here is that she holds a bouquet of carnations. These flowers are cheap too, yet I've loved carnations ever since I was little. I love their fragrance, the way they look and better yet, they are my birthday flower.

The little tag that is connected to it says "This flower reflects a desire for constancy. It is also the symbol of truth." I love the truth, I even have a pillow on my bed that reads "truth." So the January angel stays.

I guess the moral of the story is this: there's a story behind every chachka, if it's a good one, keep it. If there is true sentiment, love it, like you love the friend, family member or philosophy behind it.

Just one request: no more chachkas please. The room is full. Thank you.

Here is an e-mail from Linda to me sent in Feb. 2007:

It was so nice talking to you on Thursday, the time flew. There is still some information I would like to give you, but I have to go through my journals, and because there has been a lot of turmoil in my life unrelated to thalassemia, sometimes it's tough to look at the hurtful stuff. You almost relive it through writing and talking about it.

In a lot of ways, I believe that illness will never hurt your soul, so it's not as bad as when people hurt you, because they get your heart. The illness only gets your body. The body can become immune to things and the heart cannot, and unless we shun the whole world away, which is worse, we will always be susceptible to being heartsick.

Chapter 7

Amy's Loves

I am greeted at the door by Amy's husband Ted, and Amy, in a neat white casual outfit. She is petite, with perfect features, and the smoothest-looking skin. The house is a wood and cream-colored stucco colonial on a quiet street. It has an enormous porch with many soft chairs. Inside, the living room is filled with chairs and tables abundantly heaped with ornate metal lamps and glass vases. The dining room, beyond the living room, is dominated by a large table which is crowded to the edges with many more vases and lamps of different shapes and sizes, as well as other collectibles. They are all part of what Amy calls her "collecting and spending."

"I do too much of both, but I love it," she says.

"Where did I get all of this? Ever since my first blood transfusion, I've gotten rewards. If I get blood, I get something. My first was a toy, I forget which one now, but I've always gotten something ever since that first time, from my parents or grandparents or friends; every time I've had a treatment or a problem. I've been spoiled rotten, and still am."

I was led into the small kitchen off the dining room where Amy's mother sat on one side of a large kitchen counter, while her father sat on the other, facing her. They stayed there throughout my two-hour visit. Her mother was playing solitaire when I came and when I left.

Amy commands the house and the household. Her parents have doted on her throughout Amy's life. She tells me that they now

74

feel that "I've been too free with my spending, and I'm not saving enough, but I do what I want from day to day, now more than ever. I do what I feel like every day."

I was surprised at how well this 50 year-old patient with Cooley's anemia looked and sounded. She had an absolutely clear complexion with no discoloration from excess iron. She showed me her abdomen where she has received so many infusions of Desferal under her skin almost every day for the past 30 years, but she only had a few scars.

I could hardly believe how well she looked. I knocked on wood. She smiled. She looked more like 35, and when she talked, though her voice was small, she sounded as fresh, wholesome and vivacious as someone who never had any more severe medical or physical problems than a bad cold.

She spoke very calmly yet forcefully about her whole life and the trials of her illness. Her objective and straightforward recital of her terrible ordeals amazed me. My first reaction of wonder at her strength stayed with me throughout my visit. And remains with me now, several months later, as I tell her story.

She was born Amelia V. on October 30, 1956. She was a "colicky" child: cranky, small and sickly. She vomited frequently as an infant. She was diagnosed with Cooley's anemia at 22 months. Dr. Brancado, her hematologist, told her mother initially "either she has leukemia or Cooley's anemia. If she has leukemia, she'll be dead within a year. If it's Cooley's, she might have a little more time. Take her home and do what you can."

Her parents, both 27 at the time, confronted with these dire pronouncements about the fate of their first child, were totally distraught. But they did all that they could for her, and have continued to do so for the past 50 years. Amy says, "We are uncommonly close; it's too much. On the other hand, I don't regret it. They have given me the best life and their whole lives."

Both of her parents were born in the US, and had never heard of Cooley's anemia before Amy's diagnosis. Her father was a butcher, first in New York, and then at Lakehurst Naval Base in New Jersey. Her mother never worked after she was born. Her mother also tried not to have any more children. She had had several miscarriages before Amy. She became pregnant once "by accident" when Amy was 12, but the baby was still-born.

Amy began receiving transfusions from Dr. Brancado just before her second birthday. She had "usually a half a unit or so every few weeks when my hemoglobin fell below five," a very low level. Initially, the blood was not cross-matched. It was given, first by a cutdown (a surgical incision in a vein) at her ankle; and subsequently, by infusion into the jugular vein in her neck, a procedure that another doctor told her parents a year later "was barbaric."

Amy remembers "a nurse, Norma, who used to cover my eyes with a cloth" when the jugular vein was used. The treatment and care with Dr. Brancado was quite expensive, but her father wanted her to receive "only the best." And from the results, it proved adequate for those early days. From the time she received a first transfusion in her arm from another physician at age four, she began to have them done that way by Dr. Brancado as well.

From age five onwards, until she went to college, Amy lived with her parents and maternal grandparents in a house bought jointly by them in East Brunswick, New Jersey. There were many family conflicts, with her grandmother being particularly domineering, but everything was done with Amy's best interests in mind. Her father and mother tried to protect Amy from the emotional turmoil resulting from three generations and a sick child living in the same house.

Amy's care was transferred to Dr. Arthur Factor, a pediatrician in New Jersey at St. Peter's Hospital which was affiliated with New York Hospital. Amy loved Dr. Factor and was his patient until she went to college. She recalls with a smile that he affectionately called her "Miss Amy."

She was now transfused usually every two weeks to keep her hemoglobin above eight instead of five, and she felt somewhat better.

She remembers "breaking a leg while running at age five or six, and lots of infections and bone pain almost all the time. And I missed about 50 days of school a year because I needed to be cross-matched on Monday, and my father would take the day off, and after my cross-match, he would take me to lunch or something fun for the rest of the day."

Of course, she would need to spend another day after her cross-match getting her blood transfusions at the hospital. Nowadays, it's all done in the same day. Back then, she and her parents had a busy medical schedule almost every week of her life.

Her transfusion requirements increased at about age 16, in 1972. At that time, she was cared for by Dr. Virginia Canale, a pediatric hematologist at New York Hospital. There, she had her spleen removed. The splenectomy was complicated by painful abdominal adhesions due to spleen hemorrhages, with subsequent scarring in the area.

Amy began college at Trenton State in 1974, and moved away from home for the first time in her heretofore protected life. With this change, Amy was more physically active and had increased bone pain and shortness of breath. "I had more heart failure problems and more shortness of breath. I had transfusions less regularly, about once a month." Her hemoglobin was maintained at around seven. "I had lots of bone pains. It was so bad I could hardly walk."

Because of her tenuous medical condition and her pain, her father often visited her and drove her around campus, door-to-door to and from classes. Finally, she was forced to move back home in her senior year. Her father bought her a car then, and she remembers having trouble "even getting into it" with her bone pain.

The car made her feel "free," in that she could, at least, get around on her own. Her increased bone pain was probably due to the expansion of her marrow spaces as her body tried to make as much blood as possible, even though the blood cells she was making did her little good. And she continued to be short of breath as she was being inadequately transfused. Her heart was working overtime to pump blood around faster to oxygenate her cells, and she was running out of heart function and breath.

During Amy's last year of college, 1978, she began doing student teaching in Freehold, New Jersey. "It was a critical year for me." Her bones and heart were deteriorating even further and she became depressed. She also was having increased nosebleeds, a lifelong problem. She confided to her father then, "It's not good. I don't think I'm gonna make it."

Her father broke down in tears for one of the few times she had ever seen him do so. He said, "Amelia (he always called her that), you're gonna be OK. I promise you, you will." She managed to attend her graduation, "even though I could barely walk up to the stage to receive my degree."

And then the magic happened. It started when her friend and fellow patient, Linda Chiarella, told her about a new drug called Desferal. "You have to get Desferal," she said.

"Desferal is the magic drug that saved my life," Amy says. "I went to New York Hospital and met Dr. Markenson, the best doctor I ever had. I started to get Desferal and I was dancing by New Year's Eve of 1979."

She also had begun the "hypertransfusion" regimen and had a hemoglobin level of between nine and ten. I think her improvement was mainly due to her higher hemoglobin level with her increased transfusions. But Amy attributes most of her new energy to the Desferal which certainly prevented iron toxicity, but would not on its own give her new strength. "It was a miracle and my life got so much better," Amy remembers.

At that time, she began attending a night clinic for transfusions at New York Hospital. It was a new and unique setting for her and other thalassemia patients. The medical community at New York Hospital "was recognizing that we patients have lives outside of our disease, and it was very much welcomed. It was a revelation.

"I was 22 years old, and I finally could work during the day, and take care of my medical problems at night." She had begun teaching at a Catholic school at the time. Several patients around Amy's age and younger were transfused together in the evening at the hospital so that they could hold day jobs. It was also a community of patients socializing together.

This New York Hospital experience had its pros and cons since everybody knew everything about each other: when patients became ill and died, as well as the nice things that happened in their lives.

Amy met Danny Pizzulli, also a Cooley's anemia patient, at New York Hospital. He turned out to be "the love of my life although, initially, we didn't connect. Danny was going out with Linda C. at that time, and even though I saw him looking at me a lot, I wasn't going to get involved with him any more than as a friend, because Linda was my best friend."

Amy never developed significant breasts, and didn't see herself as much of a love object back then, but "eventually I saw that he really liked me. Then, Linda told me that she and Danny were 'just good friends,' and that he really liked me in a special way. So we started going out and we really connected. We fell in love."

Amy and Danny were married in June 1980, and Amy was very much in love and happy with him. Amy's parents were very happy for her and accepted Danny as a son. Amy says tearfully, "Linda C. died the day after Danny and I were married. I don't think it was because of Danny and me, but I felt guilty about it for a long time."

Amy got a teaching job in New Jersey, and Danny worked as a lab technician at MedPath, a medical testing laboratory. They first lived in an apartment in Eatontown and then bought a house together. Life was good.

"We went to Bermuda on our honeymoon and Danny made me rent a Moped with him. Of course I had an accident, broke a leg and was all messed up. But we were really happy." While telling me this, Amy started to cry.

"Tomorrow, it'll be 11 years since Danny died, and I still remember him. We were good together. I trusted and believed in him because he had suffered like me and knew what thalassemia was like. It was special. We thought we were invincible."

Danny and Amy were happy together for 15 years, working and being treated and loving each other and living together. In 1991, they bought a house together in Matawan, thinking of their long-term life together. It was a significant financial commitment for them, but the long, long-term was not to be.

In 1993, for reasons that she cannot explain, Amy's feelings about life, especially her personal life, changed dramatically. "I met new people who opened my eyes to new possibilities in my life. To my personal freedom. I began to start feeling really good and free. I felt sexual for the first time.

"I met my current husband, Ted, during this time, and we have had a sexual and emotionally fulfilling life together since then. Life became more than just being a good patient with a good husband. There was another world out there for me and I wanted to be part of it."

Her marriage to Danny fell apart. He got angry, as expected, and they separated. They sold the house and moved into separate apartments. She and Danny remained friends, though. Her parents were initially very angry at her, but, as usual, they came around to supporting Amy's needs and Amy's wishes.

While Amy was always compliant with her Desferal, Danny was not. "The reason was never clear. He would prepare Desferal solution for me every night and it would have been just as easy to prepare one for himself. But he just didn't do it for five or six years, despite my pleading with him. Eventually, one night he went back to it, but it was too late."

Danny was more active than Amy in the CAF. Along with Ralph Cazzetta, he was an officer in the Thalassemia Action Group (TAG), the group of teenagers and young adults with Cooley's anemia I have mentioned before, who supported one another and reached out to other patients. TAG is still a very active organization.

Amy has stayed under the public radar with respect to her disease her entire life. Her father, especially, didn't want people to know about it because he thought it would mark her as special and different, and hurt her. Amy herself has always been "low key" in public, quiet about her disease and with minimal involvement in the CAF.

Danny developed liver cancer in 1995, and he died from it the next year. Many patients with thalassemia become infected with hepatitis C virus through their blood transfusions, and can die of liver failure. Some of these patients also go on, as Danny did, to develop liver cancer.

Amy felt terribly guilty and depressed when Danny passed away. "I don't know if I would have survived emotionally if Danny had developed the liver cancer and gotten sicker and we were living together in that house. I wouldn't have the life I have or be the same person I am today.

"I don't think I could have survived taking care of Danny with his illness getting worse." She continued to see Danny during his illness, and his last words to her were: "I love you. You didn't do anything wrong."

"He still really loved me."

She cries again as she says this. She has had a hard time accepting Danny's death and the happiness of her new life with Ted, even now. She has been intermittently depressed and unhappy, "morbid," she says, from that time on, even though she loves Ted, and eventually married him.

Amy has become more religious since Danny's death. She believes in "God somewhere out there." A priest told her shortly after Danny's death: "You've got to let go of Danny or you'll lose Ted. You deserve to be happy."

"Ted is different from Danny. I tell him, 'You don't understand me. You've never had thalassemia.'" But Ted is very loving, and life in the cream-colored house with Ted and her parents and the many beautiful objects she has acquired and loves is good.

Amy continued to go to New York Hospital twice a month until four years ago. It was a long trip to the hospital from New Jersey. But it was worth going for the care of Dr. Patricia Giardina and her staff, and the camaraderie of the other thalassemia patients, friends she had made over many years.

"But now they're all gone," she says. Linda D., another long-term survivor of Cooley's anemia, whom I have also written about, told me the same thing. These two women are among the few surviving members of their age group with Cooley's anemia. The others have succumbed to heart failure, cardiac arrhythmias, and other illnesses despite optimal treatment. For convenience, Amy now has her routine care managed by Dr. Dennis Fitzgerald in Riverview, New Jersey, much closer to her home.

Amy has been a third-grade teacher for over 20 years. During this time, she has seen attitudes towards teaching, which she loves, change. "Kids need discipline. Parents today don't know how to parent. There is no discipline. Also, because the parents I deal with have money, they think everything is coming to them and their kids."

Reflecting on her own childhood, she realizes that she really needed discipline to deal with her disease and her life. "I feel like I have to be more than 110% to be equal to anyone else." Even today, reaching 50 and being a Holmdel teacher of merit, she still feels the same way.

"Only a very few people at school know I have thalassemia. This is an upscale community. They'll think of blood transfusions and AIDS, and they might think differently about me if they knew. Right now, I'm just another teacher, and that's all I want to be. I keep my illness under wraps."

Amy has had many hospitalizations over the years for infections, kidney stones, gall bladder disease, and diabetes. She finally had her gall bladder removed two years ago. She resisted the operation for many years despite admonitions from Dr. Giardina. "But she [Dr. Giardina] doesn't have thalassemia. I don't want any surgery that isn't absolutely necessary. The splenectomy was bad enough. I usually got away with Motrin and Maalox for my gall bladder attacks but then I really had a bad one. I told Ted and my parents, it was time."

She has diabetes from iron in her pancreas and she takes medicine for that. She wears hearing aids for the hearing loss which is a side effect of Desferal. She has a "blocked" carotid artery, cause unknown. She takes Motrin when she needs it for bone pain and other aches.

She has taken Paxil for intermittent depression. She says she panics when any health emergencies arise, but I think that is, to some extent at least, self-protection against medical mismanagement. Some paranoia can be a very useful survivival mechanism, especially in emergencies. I am in complete agreement with Amy's approach. Tell the doctors and nurses everything you feel and think. Illness is not a time for stoicism.

Amy lives her medical life as a patient carefully and warily and she is absolutely compliant with her treatment. "I take Desferal faithfully. It is my life," she says, although over the last few months her ferritin has been so low, below 100, that she only infuses it three days a week.

She would love to take the new oral iron chelator, Exjade, which Dr. Giardina has offered her so that she could "stop poking myself in the belly," but hasn't done so yet. She is very careful about "rocking the boat" in terms of changes in her medical care.

Amy used to feel that the life she had been dealt "was not fair. I have to do all these things. Take needles in my arm for the transfusions. Take needles in my belly for the iron." Now, she feels, "life is life," and she is grateful for what it has given her and where she is today.

She bought the house she now lives in with the insurance money she received after Danny's death. But with her collecting and other "spendaholic" activities over the past several years, "I owe more money on it now than when I bought it."

She continues to tell Ted and her parents " 'Everything is mine. The house and everything in it!' I am a control freak." Her parents and Ted appear to be subdued, at least in my presence, and Amy seems to be the main focus of everything. The one thing that is clear, however, is that she deserves her treasure trove of people and possessions as the reward for her courage and spirit.

Part Two

Understanding Thalassemia

This section describes how medical scientists, including me, have tried to solve the mysteries of the development of thalassemia. How has the disease been investigated? What answers have been obtained? What did we know way back in the 1960s about the disease? What have we found out in the last five decades? What questions about the biology and chemistry of thalassemia are still being pursued, and how? How have these advances contributed to diagnosis and new therapies?

The following chapters tell the story of how the molecular pathology of thalassemia and the gene defects that lead to the disease were discovered, and how regulation of the switch from fetal to adult hemoglobin production continues to be explored.

Chapter 8

The Big Alpha

My entire life as a physician-researcher has been intricately involved with the advances in the understanding of Cooley's anemia. My personal experience with these advances, therefore, is inevitably part and parcel of the following chapters on the unfolding developments in the scientific research that has led to the comprehensive knowledge of the disease that exists today. I write it as I lived it, as a researcher immersed in the work, which, for me, could not have been more exciting or alive.

Laboratory research has been seminal in understanding how Cooley's anemia develops, and has provided unique insights and approaches that have been applied to many other human diseases. I have been fortunate to have done my part in the research in thalassemia over the past five decades. The next five chapters in the book recount the story of my research in thalassemia, interwoven with advances in the field made by the contributions of others.

For the non-scientists, I have provided overviews at the beginning of each chapter and summaries at the end. I have also included figures that might be helpful in the Appendix. Selected relevant references to the research cited in this section of the book are also provided in the Appendix.

*   *   *

In the 1960s, scientists began to understand the molecular biology of Cooley's anemia. There were important advances in the knowledge of human hemoglobin structure and function. Sickle cell disease was shown to be due to a single base mutation in the human β globin chain that led to sickling and the disease.

Beta thalassemia, Cooley's anemia, was associated with decreased amounts of normal hemoglobin, HbA ($\alpha_2\beta_2$). However, the disease was poorly understood at the molecular level. It was known that there was a problem with the production of human β globin, but how that decreased β globin caused the severe disease was unclear. This is the story of what happened next.

*   *   *

When I started my scientific research at Columbia University in 1965, all that was known for sure about the severe anemia of patients with Cooley's anemia was that there was little or no normal hemoglobin (HbA) present in their red blood cells, and that this was due to decreased or absent human β globin production.

At that time, Dr. Vernon Ingram and others provided evidence that each of us inherit two human β globin genes, one from each parent, specifying β globin production. If you inherit two "β thalassemia genes", one from each parent, you make little or no β globin and have a severe anemia called Cooley's anemia.

I had just come to Columbia University from the National Institutes of Health (NIH), where I had been introduced to intensive laboratory research on protein synthesis for the first time, and I had fallen in love with the experience of doing research. I had come to New York to continue a career in academic medicine, primarily to study a human disease in the emerging field of molecular medicine. I wanted to find out what went wrong in a human disease at the level of our proteins and our genes. I thought that it was necessary to study these molecular events in the cells of patients with a human disease in order to answer this question.

I wanted to become a hematologist as well because I was fascinated by the blood system: how the bone marrow makes blood and what goes wrong to cause disease? Training in hematology consisted of taking care of patients with blood diseases and doing research in an area of interest; my interest was in how proteins are made in red blood cells.

The only hematology program in New York working on protein synthesis in red blood cells at that time was run by Dr. Paul Marks at Columbia. I was referred to his laboratory by Dr. Helen Ranney who was then an expert in human hemoglobin at Albert Einstein School of Medicine, with whom I initially thought I would work. However, she was working primarily on the structural aspects of human hemoglobin: how hemoglobin normally folded and was misfolded in disease. I was more interested in how hemoglobin was made (hemoglobin synthesis), so she recommended me to Marks. It was a professional relationship that served us both well over the next decade.

The first project I worked on in Marks' laboratory was to investigate the molecular basis of Cooley's anemia by studying what was different about how hemoglobin was made in the diseased cells compared to normal red blood cells.

At that time, it was becoming clear that genes specified proteins by using an intermediate molecule called messenger RNA (mRNA). While DNA is the immovable stable source of the genetic code in the chromosomes, mRNA, a complementary copy of the DNA, is used as the copy of this genetic information to provide the triplet genetic code from which proteins like globin are synthesized.

Hemoglobin DNA like all DNA is a long sequence composed of four different nucleotides or bases, adenine (A), guanine (G), cytidine (C) and thymidine (T), strung together in a certain order which is specific for the protein it encodes. For example, …CAGATGGGAACC… is part of the sequence in the hundreds of bases of human β globin DNA.

Using enzymes called RNA polymerases, the mRNA copy is made from the DNA template, thus preserving the order of nucleotides in the DNA. There is specific base affinity or pairing, between A and T, and G and C, in the double stranded DNA of the DNA helix. When mRNA is made from DNA, if A is on the DNA, uridine (U, the base in RNA instead of the T in DNA) is inserted in the mRNA; when T is on the DNA strand, A is inserted in the mRNA. Similarly, if G is on one strand of DNA, C is placed on the other strand and vice versa.

If ... CAGATGGGAACC ... is the sequence in globin-coding DNA, then its complement ... GUCUACCCUUGG ... is the sequence in globin mRNA.

The so-called "genetic code" had recently been discovered while I was working at NIH in the early 1960s. This code means that three nucleotides (triplets) or "codons" in mRNA specify one and only one specific amino acid in the growing string of amino acids that results in the final polypeptide or protein product, such as β globin. This process of protein synthesis happens on large supportive structures called ribosomes. Ribosomes are complex beds made of many structural proteins, upon which the mRNA strands rest, and the amino acids are assembled.

For example, the 12 nucleotides, GUCUACCCUUGG, in globin mRNA specify (or carry) the genetic code for four consecutive amino acids in the growing globin chain on ribosomes: the amino acid valine is specified by GUC; then tyrosine is specified by UAC; proline is encoded by CCU; and tryptophane by UGG. Special molecules that I had studied at NIH, called transfer RNAs (tRNAs), are intermediates in this amazing process that ensures that the right mRNA triplets insert the right sequence of amino acids in specific proteins, like human β globin.

The relative simplicity of the process by which proteins are made has always impressed me. Four nucleotides in DNA and RNA, specifying 20 amino acids in proteins, are the bare bones of the entire make up of most organisms, including humans. Just four elements–carbon, hydrogen, nitrogen, and oxygen–provide over 99% of the structural components involved!

What's wrong in thalassemia cells

My first project in the Marks lab was to study how the ribosomes of cells from patients with Cooley's anemia differed from those in normal red blood cells. I used a so-called "cell-free system" to do this. I assembled the components needed to synthesize a protein in a test tube instead of an intact cell. I isolated ribosomes with their particular mRNAs attached to them (polyribosome-mRNA complexes) from normal red blood cells, and from those of patients with Cooley's anemia.

While nucleated red blood cells in the bone marrow synthesize and accumulate most of the hemoglobin, a small number of red blood cells still capable of making new hemoglobin persist in circulating blood. These red cells called reticulocytes have lost their nuclei but still retain their ribosomes and globin mRNA, and, thus, their capacity for globin synthesis. I used these reticulocytes in the circulating blood of patients to do these experiments.

A major advantage of using reticulocytes is that most of the mRNA on reticulocyte ribosomes is globin mRNA because globin is the major protein that red cells make. To a test tube containing either normal or thalassemia ribosomes and their mRNA, I then added all of the other general chemical ingredients needed to make proteins, such as amino acids themselves, tRNAs, and ATP as an energy source. Since the major protein in red blood cells is globin, we thought that the polyribosome-mRNA complex in these cells would mostly reflect globin synthesis, and might reveal less β globin mRNA activity in the red cells of patients with Cooley's anemia than in normal red cells.

To my surprise and delight I found that, indeed, the polyribosome-mRNA complexes in red cells from patients with Cooley's anemia made significantly less protein than those from normal subjects. I had mimicked in a test tube (*in vitro*) what was true in intact thalassemia red cells (*in vivo*): there was less globin being produced than normal.

I also showed in these experiments that there was nothing wrong with the ribosomes themselves in the Cooley's patients' cells. I used a synthetic mRNA, polyuridylic acid or polyU, to show that the thalassemia ribosomes could support general protein synthesis as well as normal ribosomes. I had excluded the ribosomes themselves as the culprit. My results were most likely due to deficient or defective β globin mRNA in the cells of patients with Cooley's anemia.

I could not have been happier. We published these results, the first to suggest that there was a defect in β globin mRNA in Cooley's anemia, in the Journal of Clinical Investigation in 1966.

The Clegg column

Next, I wanted to prove that the decreased protein synthesis I had found using the cell-free system was really due, as I suspected, specifically to decreased β globin synthesis. I discovered that there was a new technique available to accurately separate and quantitate human globin chains, developed by Dr. John Clegg working with Drs. Mike Naughton and David Weatherall at Johns Hopkins.

The principle of the separation was to use a high concentration of the chemical, urea, to prevent different globin chains, such as the α and β globin chains in normal human HbA ($\alpha_2\beta_2$), from combining together. Without urea, the different globin chains tended to stick together in solution, and that prevented their separation and isolation. Some fetal hemoglobin (HbF, $\alpha_2\gamma_2$) is also present in normal and thalassemia blood, and so γ globin chains have to be separated from the α and β globin chains as well.

In Clegg's procedure, the total globin in the red blood cells is isolated and dissolved in urea. The globin mixture in urea is then placed on top of a vertical glass cylinder containing resin (like sand) that allows the globin chains to separate from each other as additional urea solution is passed through the column resin bed. The different globins are washed off the column (eluted) by increasing the salt (NaCl) concentration in the urea solution. As the different globin chains have different physical and chemical properties, they leave the column at different salt concentrations. First, γ globin comes off; then β globin; then α globin.

I was very interested in this new discovery and visited Hopkins as soon as I could. I was treated most generously and warmly by John Clegg and his colleagues, and learned the procedure. It was one of the most important technical advances that I have ever learned and used in my research career, and that was especially so at the time. New technology is usually the impetus for new biological insights, and the Clegg column was certainly one of these key innovations.

The most important thing about the Clegg column for me was that it not only separated the different globin chains for the first time, it also allowed the measurement of ALL of the *newly synthesized globin* being made in cells.

Clegg measured the globin chains being made in normal and thalassemia cells in circulating blood. The small number of red blood cells in circulating blood capable of making new hemoglobin, the reticulocytes, were again the important cells in these experiments.

To track the globin protein being newly made in the reticulocytes, Clegg added a mixture of amino acids containing some radioactive ones, such as leucine, which is abundant in globin protein, to the blood sample. He incubated the blood containing the reticulocytes at body temperature, for short times, usually an hour or less. All of the globin being made, both in intact HbA and HbF and also as free globin chains not in hemoglobin, could be detected by the Clegg column.

The reticulocytes incorporate the radioactive amino acids added into newly synthesized globin protein. The cells are then broken open (hemolyzed or lysed) by exposure to water, and the total globin is isolated. When the red blood cell hemolysate is dripped into a solution of ice-cold acid-acetone, the globin separates from the heme, and is precipitated while the heme remains in solution. The globin, separated from the heme, looks like newly fallen snow, clean and pure and white. I enjoyed the sight. The total globin precipitate is then dissolved in the urea-salt solution.

The urea solution containing the total globin is placed on columns of resin (five to eight inches long, and about one inch in diameter), and the chains are washed off the column with large volumes (several hundred milliliters or mls) of 8M urea-salt solution to ensure complete separation of the different globin chains. First, the γ chains come off, then the β chains, then last, the α chains. I had been the fortunate beneficiary of exposure to this technology, known as column chromatography, at the NIH. I learned the technology from Dr. Herbert Sober, a pioneer in the field with whom I had taken a course on the subject. I was falling in love with protein biochemistry then, and my human globin studies only increased my ardor for the technology.

As globin-containing urea solution dripped off (was eluted from) the bottom of the column, it was collected in glass tubes so that the amount of each globin and the radioactivity in the separated chains could be measured. The separation of the globin chains by the Clegg column is clear and reproducible and still hypnotizes me to this day, although other methods have since simplified and supplanted this tedious process.

The column samples containing the separated globin chains are first measured for their protein content using a spectrophotometer, a device for optically detecting the amount of each globin protein; this protein is mainly a result of stable intact HbA in the mature red blood cells. As expected, these protein measurements showed relatively equal amounts of α and β globin in both normal subjects and thalassemia patients. In the thalassemia patients this is because most of the red blood cells present (and globin chains in HbA) are from transfused blood.

The reticulocytes, however, only represent blood made from bone marrow production; they are low in number, but they are the only cells in the sample that synthesize new globin. The content of the newly synthesized globin chains was what I was after. I could analyze this by measuring the radioactivity in the fractions collected (or eluted) from the column. This was the key measurement for me because it told me the total amount of each globin chain being made.

To measure this globin synthesis, a milliliter (ml) or less of eluate sample was mixed with a dioxane-containing solution in small glass vials. The dioxane solution permits the radioactivity present in the sample to be measured or "counted" in a "scintillation counter," captured by the counter as light emitted. This measurement reflects the amount of each globin chain being made in the blood samples of the patients.

As expected, in samples from normal subjects, there was relatively equal amounts of radioactivity (globin synthesis) in the α and β chains. But in my very first experiments with thalassemia samples, I got an enormous and thrilling surprise. Every time I looked at the radioactivity pattern, I saw a vastly greater amount of radioactivity coming off the column in the α globin region compared to that in the β region. There was fivefold or more α globin as compared to β globin in the thalassemia samples, sometimes tenfold or twentyfold more; and in the case of no β globin being made at all, infinitely more α globin. I had discovered the big alpha (α), the vast excess of free α globin chains being made in thalassemia cells.

I knew my results were unique and meaningful, but I really didn't understand what they meant right then. I was puzzled. I had discovered that while there was no extra α globin accumulating in the protein present in thalassemia red cells, there was lots of excess α globin protein being synthesized. How did this happen? What did it mean? It took me a while to figure that out.

Life in the lab

I was a human lab rat back then, completely content in the lab doing experiments with my own hands. If you want to do science, you have to love the process of doing it, not be focused on the result. If you're only focused on results, don't become a scientist.

I was happiest when I was all alone in the lab in the evening or on a weekend, working on an experiment. I knew exactly what I wanted to do, and how to do it. I was following a sure-fire protocol: a recipe for a special meal. I felt like a master chef, except here the meal was experimental results no one had ever seen before or enjoyed in quite the same way. You can believe what you see because you've done it all yourself. It's just you and biology at work. Very little is left to chance or to others.

And in the best of times as a lab rat, I had the majesty of silence and my own thoughts as the only intruders, as I did my thing. It is an unusual and fulfilling exercise if you're hooked, and I was hooked. In the experiments themselves, no mistakes or new wrinkles are allowed. You've got to pay close attention, or you'll have a very unsatisfactory experimental meal.

No radios, earphones, phone calls allowed. Just your own thoughts. In running the Clegg columns and analyzing the results, I used to love to look at the numbers as they came off a printer attached to the scintillation counter, similar to that of a supermarket printout of charges, here indicating how much radioactivity was, first, in γ globin, then in β globin, and finally in α globin. I could draw the graph of the results in my head as they appeared on the printer output, and calculate the excess of α globin right then and there. I knew what the raw data meant immediately. It was a special thrill.

Interruptions by others were most unwelcome when I was in the lab. Everything's fine until the world intrudes. And as you get older, that happens with more responsibilities and a bigger lab. The world intrudes … and your lab rat days are over. Mine ended in the mid-'70s. From then on, I have had to live vicariously through the first hand lab experiences of others (fellows, graduate students, technicians), although to this day, I always think like a lab rat; I still always look at the primary data, and design, supervise, and oversee experiments like one.

Most of the people I have worked with in the lab over many years have also been lab rats. I use the term as the highest compliment I can give to my scientific coworkers. If you don't like being and thinking like a lab rat, you shouldn't be a scientist.

I, personally, together with several of my co-workers, most especially Joyce O'Donnell, a superb technician and a fine person, spent hundreds of hours in the '60s and '70s running Clegg columns with blood samples from patients and normal subjects and analyzing them. I was happy to do so.

I never went back to look at how much of each globin was made in the ribosome cell-free system I had used earlier. Intact cells were much more efficient at synthesizing globin than cell-free systems and the disparity in α versus β globin synthesis in the intact cells had essentially given me the answer I was looking for: there is a marked decrease or total absence of β globin and presumably β globin mRNA on the ribosomes cells of patients with Cooley's anemia.

My Clegg column results were due to the fact that while the Cooley's patients made little or no β globin because of their β globin genetic defect, α globin continued to be made in the same extremely high but normal amounts as in normal people. This had never been appreciated because it had never been seen before. As I surmised later, the excess of newly-made α globin chains were α chains without any β globin partners.

Why was the vast excess of α globin chains being synthesized in thalassemia cells missed by Clegg and his co-workers? It was because of the way they prepared their globin samples for analysis. After breaking open (hemolyzing) the red cells following their radioactive labelling of the cells, and before applying the samples to the Clegg column, they purified the hemoglobin from the hemolysate by dialysis, a filtration process that led to the loss of the free excess unstable α globin chains. These free α chains, including the radioactive ones, probably just stuck to the dialysis tubing (usually cellulose) that was used.

In their papers on the subject, Clegg and his colleagues (1965) reported a modest excess of α globin radioactivity in the reticulocytes of patients with Cooley's anemia. Most often, about twice as much α globin as β globin was seen, but no consistent vast excess of α globin as I found, even when they used the total hemolysates for analysis.

They also had different expectations about what they would find than I did. In their paper, Clegg *et al.* emphasized the hypothesis that "β globin is necessary for α globin release from the ribosomes." What Clegg *et al.* implied was that there could not be too large an excess of completed normal α globin chains released from the ribosomes in thalassemia cells because there was very little or no β globin to pull the α chains off. I think that was the concept that misled them.

This concept that "β globin is necessary for α globin release from the ribosomes" turned out to be completely wrong. I showed in my experiments then, and later, that α and β globin chains are made completely independently of each other and that, in thalassemia, α chains continue to be synthesized normally even when there are no β globin chains being made at all.

When I returned to New York and started doing my own experiments, I never purified or dialyzed anything from the total hemolysate because I saw no need to do so, and I didn't want to lose anything that was being synthesized. I also found no reason to purify the column fractions to remove the urea before measuring the radioactivity as Clegg did. Even though the scintillation vials containing the globin samples had clumps of white crud (the urea), I found that the radioactivity measured by the scintillation counter was unaffected by the urea and was preserved intact.

I had no preconceived notions of the expected results. I never do. I believe the data. But as I have said before, I was amazed to see five, ten, twentyfold more radioactivity in α globin than in β globin eluting from the Clegg column, in every blood sample from every patient with Cooley's anemia.

But I couldn't figure out what this finding meant for several months.

I asked everyone in the lab, from the top on down, to help me explain my findings, and they all insisted that my wild amount of radioactivity in the α globin peak was simply an artefact. "It's crazy; it's an artefact," that's all I heard. I would mope around the lab trying to figure out what all this α radioactivity meant. I went home and dreamt about it but I still had no fruitful insights to help me make sense of what I saw. It was a very eerie time for me. I was confused and frustrated.

Eureka

Finally, finally, Bob Debellis, an MD researcher in the lab at that time, turned on the light bulb for me. "That's the free α globin chains being made," he said. "They have no β or γ globin to combine with. They get lost from the cells over time and you usually don't see them in the blood, but they're being made for sure. They are the radioactivity you see." Those were the magic words that led me to figure out the rest. I just knew right then that that was the right explanation and I went on quickly to prove it experimentally in great detail.

The reason Clegg *et al.* didn't see the vast excess of radioactive α globin synthesis was that his treatment (dialysis) to purify the blood sample and isolate hemoglobin led to the loss of most of the free excess α globin chains because they were unstable. They were either precipitated in the dialysis bag or degraded. I finally understood what was happening. We published our results in the journal *Nature* in 1966, and we made a big splash in the globin field.

At the time of my discovery, there were evidence from Dr. Phaedon Fessas in Greece that there were free α globin peptides in so called "inclusion bodies" in Cooley's anemia cells. These peptides were found in most nucleated red cell precursors in the bone marrow of thalassemia patients, and were thought to be mainly α globin fragments, although a small amount of intact α globin was also reported.

How important these α globin remnants were, and their magnitude, was unknown at the time. My finding of the vast excess of α globin being made in all the thalassemia samples I studied led me to the idea that the Fessas inclusions were indeed directly related to the continued high level normal production of α globin by red blood cell precursors in the marrow of patients with thalassemia, and this was reflected in my studies of reticulocytes.

Free uncombined α globin chains were precipitated in these precursors, and these cells were being prematurely killed by scavenger cells in the bone marrow. The free α globin chains were also degraded in the red cells that survived to some extent, a process called proteolysis. Fessas was looking mainly at the results of precipitation and degradation of the free α chains being made.

The thalassemia red blood cells that do reach the circulating blood in Cooley's anemia patients are also defective; they have too little HbA, excess α globin, are deformed, and are preferentially destroyed by the spleen (the major site of red cell destruction in the body).

We also were the first to use the Clegg column to show that there were excess α over β globin chains synthesized in the cells of patients with β thalassemia trait, and also in patients with combinations of sickle cell and β thalassemia mutations.

In 1968, with Dr. Albert Braverman, my first post-doctoral fellow, I showed that, indeed, the amount of α globin chains being made in thalassemia cells was the same as that in normal cells, proving that the vast excess of these chains in the cells of Cooley's anemia patients was relative: just the result of continued normal α globin production, but with disastrous consequences because there were insufficient β and γ chains to combine with them.

The core molecular pathology of thalassemia was clarified: too little or no β globin was the primary problem, and secondarily, there was continued α globin production (at what were "normal" levels for normal red blood cells) that led to the tangles and the vastly increased red cell destruction, hallmarks of the disease.

In addition, in 1969, I demonstrated with Joyce O'Donnell that there was also a loss of at least some of the newly synthesized excess α globin by enzymatic proteolysis.

We had provided new concrete data that showed that the reason why most of the newly synthesized α globin chains were not seen in mature red cells in the blood of patients with Cooley's anemia was because these chains precipitated in the cells or were degraded, or the cells themselves were lost. The instability and insolubility of α globin at high concentration leads to these precipitates and to the inclusion bodies containing the α globin peptides that Fessas saw. The excess α globin tangles seen as the inclusions in thalassemia cells leads to the destruction of thalassemia nucleated red cells in the bone marrow at a high rate. This is the main cause of the severe anemia in β thalassemia.

Since 2001, Dr. Mitchell Weiss and his coworkers have added to the story of α globin metabolism by their discovery of a specific protein, α hemoglobin stabilizing protein (AHSP), that specifically binds and stabilizes free α globin. Thalassemia mice deficient in AHSP have more anemia than those mice with AHSP expression. Methods for increasing AHSP expression may lead to novel approaches to decreasing the toxicity of free α globin chains in thalassemia in the future.

From 1964 to 1966, I had made two important discoveries about the pathogenesis of β thalassemia in the Marks lab. One, I identified β globin mRNA as the main suspect for the cause of decreased β globin in β thalassemia for the first time; and two, I discovered that a vast excess of α globin was being made in thalassemia cells, and was the main reason for the destruction of red cells in the bone marrow and blood that led to the severe anemia.

It would take until the 1970s to define the specific mRNA defects in Cooley's anemia and until the 1980s to identify the specific mutations in the β globin genes that led to these events. But by the '70s, we were well on our way to making β thalassemia the poster child for defining the molecular biology of a human genetic disease.

The pathophysiology of sickle cell disease in which there is a single β globin chain mutation had been described earlier, but in this condition, the abnormal structure of the sickle β globin chain is responsible for the disease. There is a qualitative defect in human β globin. In Cooley's anemia, by contrast, there is a quantitative defect in human β globin. We were investigating what caused the decrease or absence of a structurally normal human protein, human β globin, for the first time. These studies led to the discovery of new regulatory mechanisms at the gene and mRNA levels in thalassemia that are the subjects of the next three chapters.

Antenatal Diagnosis

Our specific findings of the vast excess of α over β globin chains also led to the first antenatal diagnosis of Cooley's anemia, using molecular analysis of fetal blood after fetal blood sampling during pregnancy. The relative excess of α globin over non-α globin at various times during pregnancy became a diagnostic measure of the presence or absence of Cooley's anemia in fetal blood samples. These "α to β ratios" and "α to non-α ratios" using my methodology for measuring *all* of the globin chains from the Clegg column, became the standard for antenatal diagnosis using fetal blood sampling. Without the true measure of all of the excess α globin chains in the fetal blood samples, this method of antenatal diagnosis would not have been possible, since it is this value of the total numerical fold-excess of α globin over β and γ globin that is critical to the antenatal diagnosis. Precisely measuring all of the excess α globin is critical.

Antenatal diagnosis using more direct molecular analysis of the actual gene mutations in the β globin genes themselves became available in the late 1970s and 1980s, and rapidly and appropriately replaced fetal blood sampling as the method of choice. These newer diagnostic tests can be done on the DNA from any fetal cells present in either the amniotic fluid or chorionic villus biopsy. These cells can be obtained much more easily than fetal blood, and the analyses can be done earlier in pregnancy.

While short in the time span in which it was used, fetal blood analysis was the first reproducible and standard way of detecting Cooley's anemia antenatally. This methodology provided a new and unique option for antenatal diagnosis and termination of pregnancy used by many couples at risk for thalassemia worldwide in the late '60s and the '70s.

*   *   *

In summary, my first significant contribution to the understanding of Cooley's anemia was the finding that, along with decreased β globin, there is a huge previously unrecognized relative excess of free α globin being made in the cells of patients with β thalassemia. While little or no β globin is made, the usual high level of α globin present in normal people continues to be produced. These free α chains were largely undetected before my work because they are unstable and are rapidly destroyed in thalassemia cells. I showed for the first time that they are present in vast excess. These "free" excess α globin chains are a major feature of the disease and lead to the severe anemia (low blood count). The big α excess was really big. My work also led to the first antenatal diagnosis of Cooley's anemia.

I also did experiments that provided evidence that there was decreased human globin mRNA in thalassemia cells.

The work described in this chapter was crucial in defining the pathophysiology of β thalassemia for the first time. There are few or no normal β globin chains made; α chains continue to be made at a normal rate; there aren't enough β or γ chains to combine with them. The free α globin chains pile up in the earliest red blood cells in the marrow, form α globin tangles, and lead to the destruction of these cells, causing the severe anemia. The red cells that survive this process and get released into the blood stream are also damaged and prematurely destroyed.

The precise measurement of all of the excess α globin synthesized by the cells of Cooley's anemia patients led to the first antenatal diagnosis of the disease using fetal blood sampling.

Chapter 9

Human Globin Messages

The next big question to be answered in understanding thalassemia was: What is the cause of the low or absent β globin? By that time, in the mid-1960s, it was clear that DNA leads to RNA which in turn leads to the production of protein (DNA→mRNA→protein). Specific genes transfer their nucleic acid code to specific mRNAs, like globin mRNA, and then the specific mRNAs are translated into the amino acid sequence of proteins. This is how β globin is made.

By the late 1960s, the problem in Cooley's anemia had been defined at the protein level. There was decreased or absent human β globin protein made by these patients. That led to a relative excess of α globin protein. Both of these processes resulted in the premature destruction of red blood cells and the anemia of β thalassemia.

In the early 1970s, using new molecular probes, we and others would now look for defects in β globin mRNA that resulted in this decreased or absent β globin. These experiments would provide new insights into the molecular pathology of thalassemia, define novel defects in the generation of mature β globin mRNA in thalassemia, and serve as a model for elucidating defects in mRNA in other diseases as well.

This chapter details our work on human β globin mRNA in Cooley's anemia.

*　　*　　*

In the 1960s, we knew, in general, that DNA leads to RNA which leads to protein. However, we didn't know the specific role of β globin mRNA in thalassemia. Each of our cells contains the same amount and content of DNA, the blueprint for all proteins, including globin. But the human β globin genes are only expressed in cells destined to be red blood cells. We wouldn't want globin to be made in our eyeballs, anymore than we want eye proteins made in our red blood cells. How does this restriction of protein synthesis occur?

Messenger RNA (mRNA) provides the means by which specific genes are expressed in specific cells. The regulation of mRNA production is critical to understanding how specific proteins are produced. Messenger RNA is more fragile than DNA. It is a complementary copy of the sequence of DNA, but the chemical bonds that hold RNA together are more easily disrupted than those in DNA. Because of this fragility, the lifespan of mRNA in cells is shorter than that of DNA so that cells can only make the proteins they need, and only for the limited time they need them.

Globin mRNA is expressed in nucleated red blood cells in the bone marrow as they begin to make globin protein. Globin mRNA is a fairly long-lived mRNA. Other mRNAs are shorter-lived and are only used transiently to make specific proteins for different lengths of time. As in "Mission Impossible," different messages self-destruct at different times in the service of the mission. Of course, the life span of individual proteins also determines the protein content of cells. The life span of hemoglobin in red blood cells is extremely long, lasting months to years; while in the case of other proteins, their life span or that of their mRNAs may be minutes or even seconds.

Technology largely drives biology. The development of new techniques is critical to the discovery of new biological insights. New technologies do not ensure the development of new biology, but they are a critical springboard for new experimentation.

In the '70s, several critical new techniques were discovered that allowed an understanding of normal globin mRNA structure and function, and a full exploration of the defects in β globin mRNA in β thalassemia. Subsequently, other related techniques allowed all of the genetic defects in DNA in thalassemia to be identified. The new techniques, to be discussed in detail in this and the next two chapters, are:

1. The isolation and characterization of globin mRNA;
2. The use of enzymes that amplify (make more of) DNA and RNA, called polymerases; and
3. The discovery of viral reverse transcriptase, an enzyme that does the opposite of the normal process: instead of copying DNA into RNA, it copies RNA sequences into DNA.

In the case of globin, globin mRNA was isolated and used to make globin complementary cDNA or globin cDNA. This cDNA, made radioactive, was most important to us, since it could be used to track and measure the amount of globin mRNA in Cooley's anemia, and as a probe for finding the thalassemia mutations in the globin genes themselves.

In the '70s as well, the discovery of still another group of specific protein enzymes called restriction enzymes, special proteins that cleave or cut DNA only at limited and specific sequences, permitted an analysis of the organization of all genes including those at the human β globin gene locus. That is the subject of the next chapter in this book.

Finally, the pièce de resistance of the technological advances, still in the '70s, was the cloning of specific genes which led to their complete characterization. This, to me, was the true end point of the molecular biology that allowed Cooley's anemia to be completely defined at the molecular level. That is the subject of another chapter in this book.

Finding globin mRNA

Patients with β thalassemia can be divided into two groups based on their levels of β globin production. Those with so-called β-plus thalassemia produce some small amounts of normal β globin; others with so-called β-zero thalassemia have no detectable β globin.

Beta plus thalassemia patients are presumed to have either two β plus genes (genes with mutations that allow the continued expression of some β globin protein), or one β-plus and one β-zero gene (a gene associated with no β globin production). Beta zero thalassemia patients inherit two β-zero thalassemia gene mutations.

These different types of β thalassemia presumably resulted from different β globin gene and β globin mRNA defects. In the early 1970s, we wanted to know what these different defects were.

We knew that the structure of the β globin being made in β-plus thalassemia patients was the same as that of normal human β globin in HbA since the globin protein in these patients had been analyzed by sequencing and had the same amino acid sequence as that of normal β globin in human HbA. This made studies of what caused Cooley's anemia of great interest since it identified β thalassemia as a unique disease in which the synthesis of a structurally normal human protein, β globin, is caused by different specific and novel β globin gene mutations, by mechanisms as yet undiscovered at the time. Since there were no methods available at the time to study the globin genes themselves, we focused on defining the molecular details of the problems with β globin mRNA in β thalassemia.

As mentioned earlier, we published a paper in 1966 that showed indirectly that there was decreased β globin made by thalassemia cells in a cell-free system, and this suggested that there was decreased β globin mRNA in the disease, but we had no proof. We concluded in that paper that, "Taken together, the present evidence leads us to focus on a decreased or altered mRNA for hemoglobin A [HbA] as the molecular basis for the defect in thalassemia."

In 1969, Dr. Jerry Lingrel wrote a classic paper in which he isolated globin mRNA and showed that this RNA made globin. He demonstrated that mouse globin messenger RNA was a species of RNA of a particular size, 9 to 10 S, so-called "Svedbergs" (a sedimentation value or weight unit), and that it directed the specific synthesis of new globin protein.

Lingrel did this by adding mouse 9 to 10 S RNA to a cell-free system (one containing all other necessary components to make proteins) derived from rabbit reticulocytes. Reticulocytes are the young red cells circulating in the blood that I have mentioned earlier, that still have mRNA and ribosomes and still make globin, even though they have lost their nuclei. In his experiments, when Lingrel *et al.* added mouse mRNA to the rabbit reticulocyte cell-free system, mouse globin protein was produced as well as rabbit globin. Mouse globin could clearly be distinguished from rabbit globin by column chromatography. The new mouse globin could only have been made from the mouse globin mRNA since that was the only component in the cell-free system derived from mice.

As we were pushing along at Columbia on our studies of thalassemia, the Lingrel paper made a lasting impression on me. In 1971, following Lingrel's lead, two research groups showed that human globin mRNA from patients with Cooley's anemia, when placed in cell-free systems derived from other species, led to decreased β globin synthesis as compared to that of α globin. This work of Drs. Arthur Nienhuis and French Anderson at the NIH, and of Drs. Edward Benz and Bernard Forget at Yale, showed for the first time that the defects in β globin production in Cooley's anemia were reflected at the globin mRNA level in these patients. We confirmed these results in studies of our patients.

But although these experiments were important and indicated that there was decreased biological activity for β globin production intrinsic to the mRNA isolated from patients with Cooley's anemia, they did not shed any differential light on whether the thalassemia β globin mRNA was decreased in amount, or defective in its structure, or what the particular molecular defects were in the mRNA.

Dr. Danny Kacian and I at Columbia began to address this issue in the early 1970s. We eventually characterized the defects in the β globin mRNA in thalassemia experimentally. We were the first to show that different patients with Cooley's anemia had either a decreased amount of normal β globin mRNA or defective or abnormal β mRNA.

A new tool: globin cDNA

In our globin mRNA studies, Kacian and I depended on the use of the enzyme called reverse transcriptase (RT), a recent discovery at the time, that eventually won the Nobel Prize for two scientists, Drs. David Baltimore and Howard Temin. Fortunately for me, Dr. Sol Spiegelman's group including Kacian was at Columbia working with RT; they were studying a special class of viruses that use RT, called retroviruses, and from which the enzyme could be isolated.

Some viruses, like our cells, use DNA as their source of genetic material to make mRNA and then protein. However, retroviruses use RNA rather than DNA as their genetic material. To do this, they have the enzyme RT that allows them to convert their viral RNA into DNA. These viruses then integrate their viral DNA into our chromosomal DNA and then use our protein synthetic machinery to reproduce themselves.

We wanted to use RT for our own purposes: to make globin DNA, i.e., DNA complementary to the sequence of globin mRNA (globin cDNA) from globin mRNA. We could then use this globin cDNA, made radioactive, to track and measure the amount of globin mRNA in normal human red cells, and in thalassemia cells. To do this, we isolated RT from virus-infected chicken cells that made it. Drs. Inder Verma and David Baltimore at MIT reported making globin cDNA the same way and at about the same time that we did.

Danny Kacian was a postdoc in Spiegelman's lab who was already an expert in making RT. He was a true lab rat. I never saw him outside of the lab. He was baby-faced, blue-eyed, always there, always happy to be doing what he was doing. Kacian had made large amounts of RT from virally infected cells for other purposes.

He taught me the method of purification, and I made some RT myself. I enjoyed the purification of RT more than just as a means to an end for our experiments. I loved the process. Doing biochemical purifications was for me like playing in muddy water and sifting for gold.

I felt the same excitement for hands-on lab work as I did when I ran the Clegg columns, but my time as a lab rat was running out, as all kinds of clinical and administrative non-scientific responsibilities were increasing. Those were still my lab rat days, and some of my most fun days, working in the lab, but the sun was setting on them. It was the end of a phase, of a special time in my life.

From my earlier work, I was an expert at obtaining human blood samples and processing them to isolate the globin mRNA we needed. As a hematologist as well as a research scientist, I could get large amounts of material from thalassemia patients, mainly from those at New York Hospital. The cells we were interested in were again those reticulocytes, the few early red blood cells in circulating blood that still contained globin mRNA, even after losing their nuclei.

Messenger RNA, even in reticulocytes, is only a minor component of the total RNA in the cells. Most of the RNA is ribosomal RNA. Danny Kacian and I had to use a specific chemical trick to isolate globin mRNA from total reticulocyte RNA.

It was known by then that only messenger RNAs and not other RNAs in cells contained polyA tails, long sequences of adenine (A) at their ends. These AAAAs could be made to bind or hybridize to complementary strings of their partners in DNA, TTTTTs (oligodT), under certain specific chemical conditions.

OligodT sequences were attached to a small resin bed, again, like sand, in a vertical cylindrical column. The reticulocyte total RNA was poured on the column; only polyA mRNA was bound to the oligodT, while all of the other RNAs were washed off the column. Then, the conditions were changed so that the globin mRNA was washed off, and we had nearly pure human globin mRNA, since almost all of the mRNA in human reticulocytes is human α and β globin mRNA.

Then we used the human globin mRNA to make radioactive human globin cDNA we wanted by adding RT and radioactive (phosphorus 32-labelled) deoxynucleotides to the reaction mixture in a test tube. The radioactive cDNA could be detected easily even when present in trace amounts.

We then found the optimal chemical conditions under which the radioactive globin cDNA only bound to human globin mRNA, and to no other RNA. We had our magic probe, radioactive globin cDNA that could then be used to measure the amount of human globin mRNA in any RNA sample.

To do this, we established that the amount of binding of the radioactive cDNA to mRNA was proportional to the amount of globin mRNA in the sample. The more globin mRNA present, the more radioactivity was bound to the mRNA sample. After allowing the cDNA and globin mRNA to react with (or hybridize to) each other, and after destroying any extra cDNA that had no globin mRNA partner we quantitated the amount of radioactive cDNA bound to the globin mRNA.

This globin cDNA-mRNA hybridization system depended only on the amount of globin mRNA present and the time that the cDNA and mRNA interacted. These parameters and our procedures for quantitating human globin mRNA would serve as a model for later measurements of the hybridization of cDNA with human DNA as well. We would eventually use the globin cDNA to study human globin genes in human cells.

As controls for our measurements of globin mRNA content in cells, we showed that if we diluted a sample of RNA containing globin mRNA tenfold with water, then we would only get one-tenth as much of the radioactive cDNA binding to the RNA sample. If we used a sample of RNA that had no globin mRNA, none of the cDNA would hybridize.

This methodology was quite unique and very useful to other scientists as well, but it also had special problems. One important issue was that the tiny amounts of radioactive cDNA we were using stuck to glass and often disappeared. We first did the hybridization reactions in thin glass tubes that we sealed, and often, the radioactivity was gone when we analyzed the samples.

We discovered that we had to coat the glass tubes with silicone to prevent the variable loss of cDNA to glass. In science, as in other fields, the devil is in the details. We became the professors of globin mRNA-cDNA hybridization, and later of DNA-globin cDNA hybridization to measure globin genes.

Of course, our work with globin cDNA as a probe was only possible because almost all of the mRNA in reticulocytes is globin mRNA. This is unlike the situation in most other cell types that make many different proteins, and have many different mRNAs. In this latter situation, a similar analysis would have been impossible without purifying each of the individual mRNAs.

As I said earlier, using this technology, we were the first to show that different patients with Cooley's anemia had either a decreased amount of normal β globin mRNA, or, by default, if the amount of mRNA in these samples was "normal," presumably either defective or abnormal β mRNA. We published this work beginning in 1972.

Zeroing in on β globin mRNA

But even with the globin cDNA assay in place, we still had a major problem in accurately determining the amount of β globin mRNA in Cooley's anemia. This was because our globin cDNA, prepared from normal reticulocytes, was only one-half β globin cDNA; the other half was α globin cDNA. We initially used all of the reticulocyte-derived globin cDNA (which contains both α and β globin mRNA) to show that there was less total globin mRNA in thalassemia samples than in non-thalassemia samples.

We needed to find a method to separate and purify human β globin cDNA from human α globin cDNA to specifically quantify the amount of β globin mRNA in normal and thalassemia cells. If we could do this, then the hybridization of each cDNA to its globin mRNA would be specific, and one would serve as a control for the other. Only the β globin mRNA should be affected in the β thalassemia cells.

The way we succeeded in separating α and β globin cDNAs from each other was by using material from patients who had no α globin mRNA. We knew that fetuses with a unique condition called homozygous α thalassemia (or hydrops fetalis) had no α mRNA, and did not make any α globin chains. These fetuses were born dead due to heart failure, because they had no α globin genes, and produced no normal adult hemoglobin, HbA, $(\alpha_2\beta_2)$ or HbF $(\alpha_2\gamma_2)$ because of their lack of α globin.

We obtained samples of blood from hydrops fetuses, isolated their globin mRNA (which contained no α globin mRNA), and hybridized it to the combined radioactive α and β globin cDNA we had made from normal human reticulocytes. The cDNA radioactivity that hybridized to hydrops mRNA was β globin cDNA. The unhybridized and separable cDNA was α globin cDNA.

We then used these separated and purified α and β globin cDNAs to show clearly that normal people had relatively equal amounts of human α and β globin mRNA. Some patients with Cooley's anemia had decreased or undetectable amounts of β globin mRNA. Other patients had relatively normal amounts of β mRNA, but the mRNA was clearly defective in its structure, since these patients made little or no β globin. We first published these results in 1973, and follow-up papers on the subject later.

We were proud of our progress, but even with all of this good and hard work, we still did not know why patients had decreased or defective mRNA. These were further questions to be addressed quickly. The answers were in the human β globin genes.

A close encounter

My globin mRNA days were some of the fullest and richest of my life. My wife Rona and I had moved to New York in 1964 without much money; no financial security, and two new sons. David and Michael were born in Bethesda while I was doing research there, and they had to be supported and loved. I had finally established myself as an investigator at Columbia University because of my work on human globin and was moving up academically and in my research support. I was better off than I was in the big α days, but I was still a young academic cat on a hot tin roof, moving quickly from one experiment to another.

My laboratory space, when I had initially worked in the Marks lab in the '60s, was in an older part of Presbyterian Hospital called Vanderbilt Clinic. My office was a 10′ by 10′ room with no windows, and I sprayed the glass panel on the door with opaque white paint so no one could look in. I had privacy for as long as I needed it, and I really didn't mind the tiny space.

By 1972, I had moved into more luxurious lab space in a new building, the Black Building, at the southeast corner of 168th Street and Fort Washington Avenue, where I had my own laboratory for the first time. And I was now an Assistant Professor of Medicine. Although the new lab was fine in this new location, I didn't even have a 10′ by 10′ hidey-hole for an office. My only office was in the Department of Medicine some distance away, which was useless for a lab rat.

In those mRNA days, I was still making too little money, and I was busy running back to the lab at night to do the globin mRNA and cDNA experiments with Danny Kacian after my children went to bed. Or I would be moonlighting at medical jobs in the Columbia-Presbyterian emergency room or at Parkside Hospital in Queens to support our lifestyle, which included sending both children to private school. Rona was working as a psychologist part-time and helped support the family as well.

My responsibilities in the Department of Medicine were extensive and included working at Presbyterian Hospital by making rounds on patients on the general medical wards and, as a hematology consultant two months a year, caring for hematology patients, primarily those with sickle cell disease, in a hematology clinic once a week, and teaching in courses for medical students.

I now essentially was working in two laboratories: Kacian's lab and my own. Kacian was in Spiegelman's space in Delafield Hospital, now defunct and used for other purposes, across Fort Washington Avenue and down the hill at 165th Street. My own lab was in the Black Building where technicians, including my first, Joyce O'Donnell, were still running Clegg columns, and where a postdoctoral fellow, and others, were continuing to do globin synthesis studies.

I was shuttling between both labs almost every day for a few years.

The postdoctoral fellow at the time, supposedly, with my technicians, was minding the store for me in my new Black Building lab when I wasn't on site. The fellow was a very sweet and kind person, who was, however, sometimes almost totally wrapped up in his own thoughts. He was very politically active, and like most of the rest of us at Columbia, vehemently against the Vietnam War. He was often thinking about nobler causes rather than the details of lab operation.

Once, this led to a lab disaster. The lab's bread and butter experiments involved running the Clegg columns to measure globin synthesis. The enormous amount of urea used in the Clegg columns by the fellow and the technicians in the lab was very difficult to get into solution at the high concentration at which we used it: 8 molar (8M).

To prepare lots of 8M urea, we used 50 liter glass containers called carboys. We would put them on a small metal platform that allowed the carboy contents to be mixed by a large magnetic stirrer placed at the bottom of the carboy. The fellow was impatient with the time it took for the urea to go into solution, and decided to speed up the process by using a hot plate to both heat and stir the urea.

Urea is a chemical that takes up heat as it dissolves, thereby cooling the solution considerably. One day, the carboy, filled with urea, exploded as it was being heated and cooled at the same time. All 50 liters of 8M urea flooded over my relatively new beautiful pristine laboratory in the Black Building. We were up to our ankles, not in water, but in gunky, partially dissolved, slick, icky urea. It was a catastrophe. It took us days to clean the laboratory, and the smell of urea, reminiscent of urine, lingered until we left that lab, years later.

I never doubted my ability to be a successful scientist, physician and teacher, but I sure was having trouble filling each of my life's roles 100% in those days. I had my own lab, the Kacian lab, hematology responsibilities, and teaching duties; I was writing papers, going to meetings, being on committees, and I had my family life. I was overextended, as I would continue to be until 2005 when I finally relinquished my clinical and teaching responsibilities at Columbia-Presbyterian.

*   *   *

In summary, in the early 1970s, we investigated the defects in β globin mRNA that result in decreased or absent β globin in Cooley's anemia. We did this by using unique new molecular probes, radioactive globin cDNAs, that could accurately measure the amount of human α and β globin mRNAs. We discovered that there were decreased or absent amounts of human β globin RNA in some patients with the disease. In other patients, the amount of β globin mRNA was normal, but the structure of the β globin mRNA produced was presumed to be defective. Both types of patients have the same decreased or absent human β globin protein production. The details of these discoveries provided new insights into the molecular pathology of thalassemia, and also defined different defects at the mRNA level in the synthesis of a specific normal human protein, here, human β globin; similar situations were subsequently found in other diseases as well.

Chapter 10

Finding the Genes

By the late '70s, I was still flying high in my research life studying Cooley's anemia. And so many new things were happening so rapidly. We had only discovered human globin cDNA in 1972, and by 1976, we knew most of the details about the changes in β globin mRNA in the disease. It was either decreased or defective mRNA, and that was why there was little or no β globin. But now we wanted to know what caused the defects in the β mRNA? The answers were in the genes.

We used radioactive globin cDNA probes to measure the number of human globin genes in human DNA for the first time. We also utilized these cDNA probes to analyze human DNA structure and organization at the human β globin gene locus, as physical changes in DNA, also for the first time. To do this we did so-called Southern blotting, a new technique in which restriction enzymes cleaved DNA into pieces, and our radioactive cDNA identified the human β globin gene-containing pieces.

We also employed Southern blotting to detect the loss (deletion) of certain gene sequences at the human β globin gene locus that result in increased amounts of fetal hemoglobin. We found that the more deletion of DNA there was in the region between the human fetal and adult genes, the more fetal hemoglobin could be produced in adult life. These studies led to new hypotheses regarding human fetal (γ) to adult (β) globin switching. If we could

117

understand the regulation of expression of the human fetal globin genes and turn on fetal hemoglobin to the high levels made in fetal life in thalassemia patients as well as in the rest of us, we could cure Cooley's anemia.

This chapter is about how we utilized new technology to define changes in the structure and organization of the human γ and β globin genes.

*   *   *

How could we study the structure and organization of the human β globin genes normally and in Cooley's anemia? It was like looking for a needle in a haystack. There was this huge amount of DNA, the same in every cell (except sperm and eggs with half as much), and the human globin genes were just a tiny part of all that DNA. How could we find our globin genes, isolate them, analyze them, and finally, discover what was wrong at the core of Cooley's anemia?

Globin cDNA gave us a first specific way to look. It was a probe that we could use to find the globin genes in the same way we used it to find β globin mRNA. It provided us with an early entrée into the study of specific human genes in human DNA, and was a vital technological step.

Globin cDNA allowed us to find our needles, the human globin genes, in the haystack of total human DNA from any individual for the first time. For radioactive globin cDNA to work as a probe in the mountain of cellular DNA, we needed and found specific conditions in which the radioactive globin cDNA bound or hybridized to human β globin gene sequences only, and not to any other irrelevant DNA.

It was exciting to study human DNA. We broke open the cells, got rid of all the proteins and RNA, and then precipitated the remaining DNA in an alcohol-like solution. The long strands of chromosomal DNA are much like sperm fluid (which is largely DNA). We collected the DNA, a white stringy precipitate by spooling it onto a glass rod, by twirling the rod in our fingers. We then dissolved the DNA in salt and water for further use.

It was incredible to me then, and still is today, that we could arrange conditions in which the only human DNA that bound to our radioactive human globin cDNA were the human globin genes themselves, even in this world of unrelated DNA. We pursued the technology as fast as we could.

By 1976, we were using human globin cDNA to make actual biochemical measurements of the number of human globin genes in total human chromosomal DNA. Prior to that time, it was surmised, from genetic data from families with sickle cell globin genes or other mutant human α and β globin genes, that there were only two β globin genes and four α globin genes in the total DNA complement of most people. These were indirect estimates of our human globin gene content.

We could now try to ask the questions: Can we directly experimentally measure the number of globin genes physically present as DNA in our total chromosomal DNA? How many total human globin genes are there in this total human DNA? How many α globin genes? How many β globin genes? Were there fewer β globin genes in Cooley's anemia?

We did the experiments and got some limited answers. Again, this was possible only because of the specificity of our human α and β globin cDNA probes.

We followed the radioactive globin cDNAs to find our globin genes in chromosomal DNA. Using this technology, two superb Italian postdocs, Roberto Gambino and Francesco (Checco) Ramirez, found that there were indeed, as we suspected, very few α and β globin genes in total human DNA, less than 10. Then using the specific α and β globin cDNA probes, we determined, on the basis of our hybridization results, that there are about two to five of each. Not bad for a first quantitation of specific genes in human DNA.

The good news was that we had our first numbers, but the limitation was that we could not obtain more precise data using these methods. We couldn't tell, for example, whether there were two, three or four of either α or β globin genes in our DNA samples.

Fortunately, as fate would have it, two new technologies, restriction mapping and gene cloning, became available between 1977 and 1979 that allowed us and others to define the number of normal human globin genes and their organization, as well as the DNA defects in Cooley's anemia. By the middle of the 1980s, most of the DNA defects in thalassemia were known. It was another wonderful time in my scientific career.

Southern blotting

Restriction enzymes are a group of proteins, discovered in the early 1970s, that cut (or cleave) DNA only when there is a certain specific (restricted) sequence of deoxynucleotides (bases in DNA), usually four to six bases long, present in a row. If the restriction sites in total DNA occur infrequently, then fewer total pieces of DNA are produced by the enzyme cutting within the more than 10 trillion bases in our DNA sequence. If the DNA sequence recognized by a restriction enzyme is fewer, say four bases in a row instead of six, and occurs more often, many more DNA pieces are generated.

For example, GAATTC is the sequence in DNA that is cut by the specific restriction enzyme, EcoRI. Since it takes six bases present in this particular order to generate the cut (or cleavage site), relatively few DNA pieces are produced by EcoRI cleavage of total human DNA.

The beauty and elegance of restriction enzyme analysis is that the same pieces of DNA are always generated from a particular sample of DNA with a given enzyme. Therefore, the results are "repeatable": the pieces are always the same size when the digestion is repeated within an experiment. The results are also "reproducible": the same values can be obtained in different independent experiments. Repeatability and reproducibility are hallmarks of science.

Prior to restriction digestion, literal shaking of DNA, so-called sonication was the inexact process we used to obtain small enough pieces of DNA for analysis. Such large molecules as those of the DNA in our chromosomes can be randomly degraded by sonication. By contrast, restriction enzymes cut every piece of DNA in the same way at the same DNA sequence, and ensure that we are looking at most of the same pieces every time we analyze a specific person's DNA.

We separate the DNA fragments cut by restriction enzymes on gels made of polyacrylamide or agarose. These materials retard the movement of different-sized pieces of DNA to different extents. The big pieces migrate through the gel more slowly than the smaller ones.

The questions were: How can we use restriction enzymes to find the globin genes in the large number of restriction fragments of different sizes in our total DNA, and how can we look more closely at human globin gene organization and structure using these enzymes?

Dr. Chaim Aviv, a friend of mine from Israel, provided the first answer when he came to visit my lab in 1977. Chaim told me about a new technique developed in England by Dr. Ed Southern to look at gene organization in DNA. It became known as Southern blotting.

I remember Chaim telling me in a quick two minutes in the hall outside my lab then: "You take a DNA sample, cleave it with restriction enzymes, separate the DNA fragments generated on a gel. Then you transfer the fragments from the gel to a piece of filter paper (or nitrocellulose), dry it and then hybridize the piece of nitrocellulose to a radioactive probe like globin cDNA and look at the pieces." It was another absolute "Eureka" moment for me (I figure I've had less than 10 of those moments in my whole career, but I think I've been very lucky with that number).

With restriction enzymes and Southern blotting, the method of studying the organization of the human globin genes was clear. To this day, without looking it up I can still remember the cleavage site, GAATTC, for the enzyme we used most, EcoRI. It was a fact that dominated my research for two or three years. In these new experiments, we followed the radioactive human globin cDNA to find the restriction fragments of our chromosomal DNA that contained our human β globin genes, in healthy people and those with Cooley's anemia.

With Southern blotting, the restricted DNA, separated on gels, is transferred to nitrocellulose filters. Then, the filters are hybridized to the radioactive β globin cDNA. After washing the filters to remove any free radioactivity, the filters are placed next to pieces of x-ray film. Darkening of the x-ray film, black spots, on the resulting so-called autoradiogram identifies the specific DNA restriction fragments containing the human globin genes, even in the presence of restriction fragments from all of the other human DNA.

We began these experiments ASAP after Chaim told us about Southern blotting. By using different restriction enzymes alone and in combination with each other, we could figure out how the DNA pieces containing the human β globin genes fit together, like pieces of a jigsaw puzzle.

To do this, we compared the size of the radioactive pieces of total DNA that hybridized to our β globin cDNA probe when we cut the DNA with one restriction enzyme alone to what happened when a second restriction enzyme was also used in the reaction. What β globin-hybridizing fragments of DNA generated by enzyme 1 alone remained the same when enzyme 2 was added? If a piece of β globin-hybridizing DNA on the gel (and nitrocellulose) cut with enzyme 1 remained the same size after it was also cut with restriction enzyme 2, that meant that enzyme 2 didn't have a restriction site (or DNA sequence) in the piece generated by enzyme 1.

On the other hand, if restriction enzyme 2 led to the disappearance of the piece of DNA generated by enzyme 1, then the DNA sequence cleaved by enzyme 2 was present in the piece generated by enzyme 1. You get the idea. Like jigsaw puzzle pieces, the size of pieces found were compared, using many different restriction enzymes alone and together; the pieces could be linked together in a linear array or "map." Beta globin gene-containing piece 1-space-β gene piece 2-space etc.

Using enough restriction enzymes, we had the first "restriction map" of the human β globin genes at its place (or locus) as it exists in human chromosomal DNA. With this mapping, we were the first to show, at about the same time as Dr. Richard Flavell and his colleagues did in England, how the pieces of DNA at the human β globin locus are organized.

Southern blotting of the human β globin locus became our baby and we exploited it as quickly as possible to uncover the secrets of the human β globin locus and its organization. We used DNA from circulating blood in almost all of these experiments. The DNA in the nuclei of white blood cells is the same as all the other DNA in our body, and, thus, is a convenient source of DNA for analysis.

In these studies, Dr. Gregory Mears was the man of the hour. He had his own set of lab benches, known as a bay, in my lab in room 1614 of the Armand Hammer Building. It was a modern new building into which we had moved in 1975, and it gave us plenty of space to do our complicated molecular biology.

And the grant money was flowing in to keep us happy. I have always said that laboratory research is hard enough to do without worrying about the money needed to do it. I am gratefully one of the relatively few researchers who has had the good fortune to have this situation continue uninterrupted throughout my entire research career (including today).

Greg Mears was a postdoctoral fellow in hematology, doing laboratory research while still fulfilling his requirements to become a clinical hematologist. With increasing clinical responsibilities, it was more difficult to do that in the late 1970s than in my day in the 1960s, but Greg was the man.

Tall, handsome, easy-going, committed to success in medicine and science, Greg was a solid laboratory citizen: precise, dedicated, intelligent. He worked long and hard to get Southern blotting to finally work for us.

"I can't see the bands. It's too much. It's not sensitive enough." I still hear it in my nightmares today. The restriction fragments containing the globin genes were very difficult to visualize on our first Southern blots. Greg's initial frustrations with our inability to consistently see the globin DNA bands that hybridized to our β globin cDNA on Southern blots were echoed many times over the next few months as he tried to make the Southern blotting technology work: to find those specific radioactive β globin cDNA-hybridizing needles in the haystack of human cellular DNA.

I forget what technical tricks we eventually used to finally carry the day: maybe hotter (more radioactive) cDNA probes for more sensitivity, or more human DNA on the blots, but finally, Greg had Southern blotting working reproducibly and well. Most of the time, with imagination and diligence, you can find the appropriate technical tricks to make a procedure work. I and, more importantly, my superb co-workers throughout my career have been good at that.

Greg was one of them. He went on to use the technique to delineate the physical organization of the normal human β globin genes in human chromosomal DNA.

More Genes Yet

Mapping the human β globin genes at the human β globin locus, however, was complicated by the fact that even though there is only one β globin gene on each DNA strand (two per person), there is another structurally related gene, the human delta (δ) globin gene, nearby (Fig. 1, Appendix).

The δ globin gene leads to the production of the minor hemoglobin, HbA2, ($\alpha_2\delta_2$), that I have mentioned earlier in the book, whose function is still unknown. The amino acid sequence of human δ globin is very similar to that of β globin, so we expected that the δ globin DNA sequences would be very close in structure to those of the β globin gene, and that our radioactive β globin cDNA would hybridize to δ as well as β globin gene sequences. Our problem was how to distinguish the δ and β globin gene DNA pieces from each other.

Patients with an unusual hemoglobin called Lepore hemoglobin helped us. The Lepore globin protein or chain is a single globin protein that has δ globin amino acid sequences in its left half and β globin sequences in its right half in a single globin chain! The Lepore ($\delta\beta$) globin protein replaces the normal human δ and β globin proteins in patients homozygous for Lepore (with two Lepore genes). These patients have no intact δ or β globin genes.

How could this have happened? How did Lepore globin arise? It was most likely the result of an event called an "unequal crossing over" or recombination; an exchange of DNA between two DNA strands lying next to each other. This exchange of DNA is not unusual in normal sperm or egg cells, and occurs in a process called meiosis, in which new genetic changes are acquired.

In the "unequal crossing over process" that generates the Lepore globin gene, two pieces of DNA, each containing normal δ and β globin genes, line up parallel to each other, but asymmetrically, with the δ globin gene on one strand lying next to the β globin gene on the other (their finding each other is not that unusual because their DNA sequences are so similar). Then they exchange their DNA: the δ globin gene sequence on one strand moves its DNA to join the β globin DNA on the other strand, and *vice-versa*. The genetic recombination between the two asymmetrical strands leads to the formation of the Lepore ($\delta\beta$) gene.

With these genetic facts in mind, we suspected that on Southern blots we would find predictable differences, deletions, in the human δ and β globin genes in the DNA from patients homozygous for hemoglobin Lepore, as compared to blots from people with normal globin genes. To do this experiment, we searched for and finally obtained DNA from two patients homozygous for hemoglobin Lepore. They only had Lepore DNA, no normal β or δ globin gene DNA.

The key experiment was to compare the Southern blots from people with normal globin genes with those of patients with only the Lepore, the $\delta\beta$ fusion globin gene. Greg did the blots using the enzyme EcoRI. He had previously shown that when normal human DNA is cut with EcoRI, four bands containing the δ and β globin genes were seen reproducibly on the autoradiograms hybridizing to our radioactive human β globin cDNA. Just four bands out of the many thousands of DNA pieces present.

When he did the same experiment using DNA from the Lepore patients, only two β globin cDNA-hybridizing bands were seen. Two bands were missing! It confirmed for the first time, at the DNA level, that pieces of the normal human δ and β globin genes, as suspected, were deleted in Lepore DNA.

Wow! This was another powerful emotional moment for me. I still remember that day. Staring at the X-ray film on an X-ray view box, showing two β globin bands in Lepore DNA, and four in normal human DNA. What a thrill! As thrilling as any for me. I even brought home the x-ray film to show my wife and my children, a rarity for me. I also hugged Greg that day, another rarity.

The Lepore result showed, for the first time, that δ and β globin gene region sequences were missing in Lepore DNA. It was also the first demonstration using restriction analysis that human DNA sequences for a specific human gene were physically deleted (seen as specific DNA fragments), and that the deleted material could be detected in total human DNA!

The demonstration of these physical changes in DNA experimentally was the gateway to using Southern blotting to identify mutations in specific pieces of human DNA in other human genes in human disease. It also showed that we could use Southern blotting for antenatal diagnosis using fetal DNA to detect gene mutations like Lepore in humans. Quite a leap forward!

Up to that time, we did not know which of the four EcoRI bands we saw in normal human DNA with our β globin cDNA probe were δ and which were β globin gene fragments. What the Lepore experiments showed was that the two EcoRI DNA pieces that were present in both Lepore and normal DNA were obviously the ones representing the left part of the δ gene and the right part of the β gene that were present in the Lepore DNA. And the two missing pieces were those containing the DNA sequences between the δ and β genes that were lost in the Lepore unequal crossover. From the sizes of the missing restriction fragments, we could calculate the distance between the normal human δ and β globin genes.

Further experiments with other restriction enzymes clarified and confirmed the order of the pieces, and established a clear restriction map of the human β globin gene locus. The results were published in the journal, *Cell*, and this body of work was one of my most satisfying research achievements.

By 1978, we had used the restriction enzyme technology and blood samples from thalassemia patients to find out a great deal about the human β globin locus and the organization of the δ and β globin genes. However, it was difficult to find most single base mutations causing Cooley's anemia using this technology because the restriction enzyme used would have to recognize a single base mutation that either abolishes a cleavage site or creates a new one.

In 1981, Dr. Michael Baird, a postdoc in the lab, discovered such a situation in which a restriction enzyme, HphI, detected a single base change leading to a β thalassemia mutation in DNA

Drs. Mears, Bank and Ramirez in the late 1970s in the lab.

samples from several Italian and Iranian patients. This was the first time Southern blotting was successfully used to detect a single base change in thalassemia. This discovery could then be used for ante-natal diagnosis in families with this mutation by restriction analysis of fetal DNA.

Regulating fetal hemoglobin

At the same time that we were looking at Lepore and normal DNA, we were continuing to pursue a major problem that we had worked on earlier in my career: What controls human fetal globin production? As I've mentioned before, if we could just understand the mechanism(s) by which fetal hemoglobin is essentially shut off in late fetal life, and if we could reactivate it fully in children or adults with Cooley's anemia, we could cure the disease.

From the early 1970s, I knew this approach was a possibility for curing the disease, because of the existence of a condition called hereditary persistence of fetal hemoglobin (HPFH). Patients with one form of HPFH in Africa have absolutely no δ or β globin, and have no anemia. They are healthy and the only hemoglobin they produce is fetal hemoglobin ($\alpha_2\gamma_2$). They have no HbA, no β globin, no δ globin, and yet, they make enough fetal hemoglobin to be well. This is in stark contrast to patients with Cooley's anemia who have intact δ and β globin genes, and, who, despite having normal fetal globin genes, cannot make nearly enough fetal hemoglobin in adult life to avoid being anemic.

What was happening? How did people with HPFH survive normally although they never switched to produce HbA ($\alpha_2\beta_2$) or HbA2($\alpha_2\delta_2$)? They had no β globin or δ globin. The suggestion was that perhaps they had no β or δ globin genes at all!

In the mid-1970s, with the availability of β globin cDNA, we had acquired and tested the DNA from patients with an African form of "deletion" HPFH to measure their total β (β plus δ) globin gene content, before Southern blotting was available, with a collaborator from

Ghana, Dr. Felix Konotey-Ohulu. We found in these relatively crude studies, using sonicated (not restricted) DNA, results consistent with the absence of human δ and β globin genes: there was a marked decrease in β globin-like (the mix of δ and β) globin gene material in the total human cellular DNA from these patients, using our radioactive β cDNA probe, as compared to normals.

Now we wanted to use Southern blotting to see more precisely whether restriction fragments containing the human δ and β globin gene sequences were altered or deleted in HPFH, and were modified or gone on Southern blots. And, indeed, they were!

Our hypothesis regarding fetal hemoglobin regulation as we approached these experiments was that DNA in the region between the γ and δ genes (the γ-δ intergenic region), and proteins bound to these γ-δ intergenic sequences, controlled the amount of γ globin that could be expressed (Figs. 1, 2, Appendix). In people with normal globin genes and in most patients with β thalassemia, the γ-δ intergenic region DNA sequences are preserved, and have a role in shutting down the fetal globin genes in late fetal life. We hypothesized that deletion of DNA sequences in the γ-δ region, as in HPFH, derepressed γ globin gene expression and increased γ globin production.

In the late '70s, Southern blots enabled us to address this issue more directly using restricted DNA from people with normal globin genes, and those with Cooley's anemia or HPFH. Again, Greg Mears did the Southern blots, and his results confirmed and extended the observations we had made using total sonicated DNA. People with normal β globin gene loci and those with Cooley's anemia had all four of the δ and β globin gene EcoRI fragments while people with HPFH had none of them. Again, we had discovered something new about globin biology in showing the physical loss of specific human DNA sequences at the β globin gene locus associated with increased fetal hemoglobin output for the first time. We published these results in 1978 as well.

We were also interested in comparing the DNA defects of HPFH patients with those of certain Italian patients who had another disease syndrome, called δ-β thalassemia. In contrast to the situation in patients with HPFH who have no anemia, the patients with δ-β thalassemia have some anemia, but still much less anemia than patients with Cooley's anemia. Most interestingly, in homozygous δ-β thalassemia patients, like those with deletion HPFH, there is no human δ or β globin produced, no HbA or HbA2.

We, therefore, looked at possible correlations between the amount of DNA lost in the γ-δ intergenic region, on the one hand, and the amount of γ globin gene and HbF compensation on the other. We found two Italian patients with homozygous δ-β thalassemia through collaborators from Sicily. Using DNA from these patients, Greg showed by Southern blots that they, like the African HPFH patients, had deletions of some δ and β globin fragments, but, in contrast to HPFH, they also retained some δ globin gene material. They had less deletion of δ and β gene sequences than the HPFH patients and less γ globin compensation than in HPFH, and were anemic!

Thus, we saw a correlation between the amount of deletion of gene material in the human γ-δ intergenic gene region and the amount of fetal hemoglobin production possible in adult life: the greater the deletion of genes in this region, the more γ globin that could be expressed, and the less the anemia. On the one hand, in HPFH, full deletion leads to full human γ expression. In δ-β thalassemia, somewhat less deletion leads to somewhat less γ globin and some anemia. On the other hand, when there is no deletion, as in most patients with Cooley's anemia, much less γ compensation is possible, and severe anemia is the result. Our results were consistent with the idea that the γ-δ intergenic region is a powerful regulatory region determining γ globin gene expression and HbF production in adult red blood cells (Fig. 2, Appendix).

We published all of these results, and included them in a review article that Mears, Ramirez, and I co-authored in *Science* in 1980. The different relative extents of the deletions in HPFH and δ-β thalassemia we found in our experiments using total human DNA and restriction analysis were confirmed a few years later when the entire

β globin locus from these patients was cloned, and the complete sequence of the γ-δ-β globin region in these patients was reported.

We were convinced of the role of the γ-δ intergenic region in the control of human γ globin gene expression in 1980, as we are today.

In the late '80s, however, data consistent with a different hypothesis for the regulation of γ globin gene expression emerged. Dr. Bernard Forget and his co-workers published articles that suggested a completely different reason to explain why HPFH patients had high fetal hemoglobin.

In these studies, Forget *et al.* provided evidence that sequences downstream (to the right) of the deletion in HPFH that were joined to the human γ gene upstream were responsible for increasing γ globin gene expression. They showed experimentally that these downstream sequences contained so called "enhancer sequences," sequences known to stimulate the expression of genes with which they are associated. It was suggested that these enhancer sequences were brought into the vicinity of the γ globin genes by the deletions downstream, and that their enhancer effects increased γ globin expression in HPFH; and that it was these enhancer effects, and not deletion of the γ-δ intergenic region *per se* and its consequences that did so, as our data suggested.

However, several subsequent studies have been published since then that validate our explanation: first, smaller deletions of sequences 3' (downstream) to the γ gene in δ-β thalassemia and other cases of HPFH, do not bring Forget's 3' enhancers into the region of the γ gene and these deletions are still associated with increased fetal hemoglobin. The Forget enhancer hypothesis cannot explain the increased γ globin expression in these cases. Also, there has been no subsequent confirmation of Forget's enhancer hypothesis that I have seen published in recent years. Additionally, more recently, Dr. Peter Fraser has shown that much smaller deletions of the human γ-δ intergenic region are also associated with increased γ globin production, again, with no enhancer sequences in sight.

We have continued working on the details of how the specific regulatory sequences in the γ-δ globin intergenic region affect γ to

β globin switching, and, over the past 15 years, have gained new insights into the process. A later chapter discusses these experiments.

We have been the only laboratory working continuously on human globin gene regulation and talking about the role of the human intergenic γ-δ globin sequences in human hemoglobin switching for more than 30 years. However, the question of the extent to which human γ-δ intergenic sequences control human γ globin gene expression still remains unresolved.

Science, like most other fields, is a contact sport when it comes to competitors in the field acknowledging the new and exciting results of others, and assigning credit for them. The competition is intense and the acceptance of others' contributions is sometimes slow and grudging. Validation and recognition of significant results are largely represented by grants and publications, and there is a real and serious competition for those.

Of course, most of all, I love the process of doing science, the freedom of being an independent scientist, and the excitement of new discoveries that result from my research. Practically, however, what is most important to scientists like me is to have the money to do the experiments, and to have the work published. I can only be grateful that most NIH study sections reviewing my grants, and journals reviewing my papers, have given me continuous encouragement for over 40 years.

An over-riding redemptive aspect of science is that it carries the drumbeat of truth throughout its length and breadth. If the work is right, the science will eventually be recognized by the larger community of scientists.

*   *   *

In summary, during the 1970s, we used our new molecular probe, radioactive globin cDNA, to discover the organization of the human β and δ globin genes. We described the specific physical pieces of DNA involved in their structure by the use of restriction mapping made possible by a new powerful technique called Southern blotting. We showed physical deletion of human DNA between the human δ and β globin genes in certain disorders of human hemoglobin, using this technology as well.

We also found, using Southern blotting, that deletions in certain gene sequences at the β globin gene locus is associated with increased amounts of fetal hemoglobin. We identified a region of DNA between the human γ and δ globin genes as being active in γ globin activation. Deletions in this region increased fetal hemoglobin in adult cells. If we could turn on fetal hemoglobin to high levels, we could cure Cooley's anemia.

Using Southern blotting, we found one single point mutation that caused Cooley's anemia. This was the first time Southern blotting was successfully used to detect a thalassemia mutation. Globin gene cloning would be required to identify most of the other point mutations. That is the subject of the next chapter.

Chapter 11

The Ultimate Answers

In the late 1970s, a new technology called gene cloning became available that allowed scientists to separate and microscopically analyze every piece of our DNA. Any small piece of our DNA (one part of millions) could be isolated, purified, and analyzed for its structure.

We and others utilized this remarkable method to isolate and characterize normal human β globin genes and those from patients with Cooley's anemia. This chapter describes these studies that resulted in the discovery of many different single base mutations in the β globin gene that cause thalassemia; these mutations explain the decreased or abnormal β globin mRNA, and the concomitant decrease or absence of human β globin.

The identification of these mutations has also led to more precise and available antenatal diagnosis of thalassemia. These studies also served as a model for determining the DNA defects in other human genes identified subsequently that cause many other human diseases.

*   *   *

Cloning globin genes

In 1978, the extraordinary field of gene cloning was in its infancy, and we were there, in on the ground floor, to utilize this exciting new technology. Genes could now be isolated, sequenced, and characterized as never before. With this most powerful technology, it was now possible that any human gene could be cloned, its DNA sequence obtained, and its precise functions studied. We could now find the ultimate answers to the question: what is wrong with the human β globin genes in Cooley's anemia?

Gene cloning permitted what restriction enzyme analysis did not: the precise determination of the single base changes in DNA that are the most frequent cause of Cooley's anemia. We could now know precisely what specific mutations were in the β globin genes that explained the abnormal function of these genes at the level of β globin mRNA and β globin that we had been studying for decades.

To clone individual genes, total DNA containing all the genetic material is isolated, cut by a restriction enzyme, and each piece of DNA is chemically manipulated and put into separate individual bacteria (usually benign *E. Coli*). Each bacteria, containing a single piece of human chromosomal DNA, is then grown and analyzed for its nucleotide sequence by using one of several methods of DNA sequencing.

The usual procedure in the early days of gene cloning was to obtain a clone of human DNA in bacteria, grow each of the individual bacteria containing the human DNA to a large volume, and then sequence the human DNA in the bacteria; and from the nucleotide sequence, try to determine what protein the gene encoded. This was very much a fishing expedition, because much of the DNA does not code for proteins; it has other, often still unknown, functions. In the case of human DNA, the sequencing of all the clones generated was an even more off-putting exercise experimentally because there was so much DNA to characterize.

With the availability of our specific radioactive β globin cDNA probe, we had a tremendous advantage in our cloning experiments in these early days. We could find that needle in the haystack that we were looking for; we could identify the few bacteria containing our specific human β globin gene sequences with our cDNA probe, amidst the vast numbers of bacteria containing unrelated non-globin

DNA. Cloning was the method of choice for us in our quest for answers to the puzzle of β thalassemia. The β globin cDNA is like a magic fishhook that only catches the β globin gene fish, and ignores all the other ones.

Since very few such specific probes for other specific human genes were available at that time, our human β globin cDNA probe put us in a unique position. Of course, eventually, the sequence of all of the cloned DNA pieces of our globin genes and other genes would be available as probes, but that seemed a long way off in 1978. If you put different pieces of DNA in different bacteria and simply determine their DNA sequence, so called "shot-gun" cloning, it is like trying to decipher a dictionary of strange words, but with no definitions or meanings. You end up knowing the structure but not the function of the cloned DNA.

We, on the other hand, had a new way to find our word and its definition at the same time by using the human β globin cDNA probe to find our human β globin genes in the dictionary of human DNA. We could rapidly identify all of the human DNA clones containing human β globin gene sequences, and then by sequencing the nucleotides, we could identify specific mutations in the β globin genes of thalassemia patients.

Then, we could also analyze how the cloned thalassemia genes themselves differed in their expression from normal human β globin genes in experimental systems, by recreating the process of human protein synthesis, and precisely defining the reasons for the decreased or absent human β globin in individual patients with Cooley's anemia.

Together with other labs with similar interests and different patients, we spent the next several years using gene cloning to investigate the structure and function of genes and, found more than 150 different single base mutations, each of which causes Cooley's anemia.

We were excited about cloning human globin genes in 1978 as soon as we heard about the new technology because we had extensive experience using radioactive globin cDNA as a probe, and with the hybridization technology we had perfected from our Southern blotting work.

Dr. Tom Maniatis and his associates at Caltech described the first human cloned DNA collection, known as a DNA library, containing the clones of all the genes in humans. I had known Tom and noted his cloning efforts for several years as he had traveled from Harvard to Cold Spring Harbor and then to California. His work was outstanding, the best in the field, and I had closely followed its progress. Political and religious sentiment in Cambridge, Massachusetts, in earlier days by groups against "cloning people" (even just their DNA in thousands of pieces), had, in part, driven him away from Harvard.

To learn about his work first-hand, Greg Mears and I visited Tom's laboratory at Caltech in December 1978 while attending the American Society of Hematology meeting in San Diego. We drove up to Pasadena on a bright light sunny day to find out about his cloning methods, and to reciprocate by sharing our restriction data on the human β globin locus with him.

Tom is a tall, thin, serious-looking man. He is hard to read by his external appearance but he has a quick, sincere smile and relates very easily to others. Greg and I sat down in his rather small dark office off his lab. I sat facing the window looking out over the green and brick Caltech campus.

Tom sat down across from me. I had met Tom before and introduced Greg to him. Tom looked very serious and sad and said to me, "We lost the library last night."

I looked out the window, concerned, and said, "Was there a fire?" I looked for smoke and destruction.

Tom said, "No, no, we lost the human DNA library."

I had never heard the term "library" used for a cloned DNA library before. Few other people had either. It is called a library because it contains all of the individual books (our genes) that can be read individually. Tom had looked so concerned about the loss he reported, I was sure a building had burned down. Tom laughed at my misinterpretation. He then told us about how his most recent DNA library had become contaminated and was thrown away, and that a new library was being made.

He described the details of producing the human DNA library and offered us his next cloned library for our use. We provided him with our novel restriction map data of the human β globin gene locus enabling him and his lab to figure out more quickly than they otherwise would have which of his bacterial clones contained which fragments of the human δ and β globin genes.

Our restriction maps were useful because we knew the size of the EcoRI restriction fragments that contained the different pieces of these genes. Tom's lab could then use this information to determine which of these EcoRI pieces were in any of the individual human β globin gene-containing clones they isolated. They could then sequence those clones to find out their normal structure, and only analyze other β globin gene positive clones containing different or overlapping EcoRI fragments.

This was the same strategy we would use, subsequently, in cloning and sequencing human β globin genes from thalassemia patients.

Our data were also important to Maniatis' lab because they provided a restriction fragment map linking clones together in the human δ and β globin region. Our map allowed subsequent EcoRI clones with sequences overlapping our EcoRI map to be oriented upstream and downstream of our pieces. Clones with sequences in common with already sequenced clones were obviously next to each other in chromosomal DNA. Thus, the physical linkage of the globin genes and their intergenic sequences in different bacterial clones could be established linearly in both directions (to the left and right of each other).

The cloning cookbook: a difficult recipe

In Maniatis' cloning protocol which followed, total chromosomal human DNA is isolated from an individual's cells, and then cut with the restriction enzyme EcoRI (you remember, cutting only at GAATTC sequences). However, unlike in usual restriction analysis, the cleavage conditions in making a DNA library are purposely arranged so that the enzyme does not cleave the DNA at all the EcoRI cleavage sites. EcoRI is only permitted to partially digest the cellular DNA. This ensures that overlapping EcoRI fragments will be isolated in different clones and can be physically linked together if they are found to have particular EcoRI fragments in common.

Since only some of the EcoRI sites are cut and others remain intact, this also means that fewer total DNA clones are obtained than would be the case with total EcoRI cleavage of the DNA. And so we had a library of bacteria with bigger fragments than we saw with our restriction mapping of human DNA, any one of which might have by chance, either none or one, or two, or three of the EcoRI pieces containing human δ and/or β globin genes.

This library, thus, had two important advantages. First, there were fewer total clones of bacteria to be screened than a library with smaller fragments. And secondly, since DNA fragments containing two or more EcoRI pieces may be cloned into a single bacteria, one could link up the order of the pieces in a linear array if there were EcoRI pieces in common between two clones.

This was a brilliant strategy and was initially important in order to link the different pieces of DNA with each other at the β globin locus. Of course, after the whole human β globin locus was cloned and sequenced and the linkage of the different human β globin locus clones established, Maniatis' partial digestion strategy was not as important. By then, we could use specific radioactively-labelled cloned DNA sequences from a known region of the β globin locus to identify other clones that contained the same or linked regions. For example, we could use a radioactively-labelled clone containing sequences unique to the left hand (5′) region of the normal human β globin gene to find clones in our thalassemia DNA libraries that had the same 5′ β globin region.

The steps in the actual Maniatis cloning procedure are considerably more complex than just partial EcoRI digestion and analysis of the DNA alone. First, it is important to have a large amount of the human cloned DNA produced in the individual bacteria contained in the human DNA library, and Maniatis' procedure made that possible. To do this, following partial digestion with EcoRI, each different piece of human DNA is joined to (spliced into) virus-like DNA particles that normally inhabit bacteria, called phage. In this process, the phage DNA is chemically linked to the human DNA fragments to form so-called "recombinant bacteriophage DNA," which is then introduced into bacteria.

The phage are used because, as with normal phage DNA, recombinant bacteriophage DNA can be amplified (increased in copy number) to high amounts after the phage are infected into the bacteria and the bacteria are grown. Growth of the phage will eventually break open (or lyse) the bacteria and form plaques which contain large amounts of the recombinant phage DNA (including the human gene DNA).

To initiate this process of obtaining large amounts of the cloned human DNA, the library of recombinant bacteriophage DNA containing human gene fragments is infected (added physically) into a growing culture of bacteria under conditions in which not more than one recombinant bacteriophage is in any one bacteria; this ensures the cloning of individual pieces of human DNA in separate bacteria.

The recombinant bacteria are spread on agar plates and grown in an incubator. Each bacteria containing a single piece of human DNA is itself amplified (multiplied in amount) a million or more fold along with its human DNA during this incubation. Each bacteria is a factory making large amounts of a single human DNA fragment as part of a cloned human DNA library. The phage then lyse the bacteria and form plaques which contain our human DNA clones. Amazing!

Our radioactive β globin cDNA probe is then used to find the few plaques that contain our human β globin genes. The plaques with the gene of interest, in our case, human β globin gene pieces, are obtained by processes analogous to Southern blotting: transferring the cloned bacterial DNA in the plaque to nitrocellulose filters, and then hybridizing the filters to our radioactive globin cDNA for detection.

Another powerful novel technical procedure, known to bacteriologists for decades, called "duplicate plating," is critical for the identification and isolation of positive β globin gene-containing clones. After infection and bacterial lysis, the recombinant phage plaques on the agar plates expose enough of their DNA to be transferred to nitrocellulose filters. To do this, the agar plate is pressed onto two successive nitrocellulose filters (in duplicate), transferring the phage DNA from the plate to each filter; each time in the same way, to maintain the individual phage plaque DNA in the same relative position on the filters as they are on the plate.

The relative position (or orientation) of the plaques on the duplicate nitrocellulose filters are marked by pinpricks on each of the filters, and similar markers identify the relative positions of individual plaques on the bacterial plate. Then, each of the duplicate filters is hybridized to the radioactive β globin gene cDNA probe in order to locate the rare bacteriophage plaques that contain our β globin gene or pieces of it.

The two duplicate hybridized filters are then each exposed to x-ray film. If we had a real clone containing a human β globin gene piece, then each of the x-ray films (autoradiograms) would have the same black spot (the signal from the hybridized radioactive globin cDNA) in the same relative position on each of the two autoradiograms.

There were many false positive dots often seen on the autoradiogram of only one filter. But if the same spot was seen in the same place on both autoradiograms overlying each other as the two films were aligned, we usually had a β globin gene clone.

I still remember the thrill of looking at the autoradiograms for the tiny black dots indicating a positive spot, lining them up to orient them with each other, and the excitement of seeing a new β globin gene-containing clone from a patient with Cooley's anemia.

After locating the position of the positive dots on the duplicate filters, we then went back to the original bacterial agar plate to find the clone; the actual bacteriophage plaque we wanted was on the agar plate that we had saved in the refrigerator. When we located it, we scooped up the phage plaques in the region of the bacterial plate that corresponded to that of the positive dots on the filter, using a spatula (small spoon).

We then set about purifying the recombinant bacterio-phage containing our human β globin gene DNA. We took the phage plaques from the region of the positive dot, infected this phage DNA into new bacteria, and spread these bacteria on a new agar plate. Then, we repeated the entire process: grow the bacteria; have the phage lyse the bacteria; transfer the phage plaque DNA to two duplicate nitrocellulose filters; hybridize the filters with the β globin cDNA probe; identify positive duplicate spots on autoradiograms (by this time we should have many more cDNA-positive dots); and again, pick the relevant plaques off the agar plate.

We would repeat the entire process over and over again, usually three to five times, until finally, we would have a filter on which all of the plaques are positive on the autoradiogram: they all light up with the β globin cDNA probe. Then we would have a pure clone and, if we are using a library from a thalassemia patient, we could sequence its DNA to find its β globin mutation. It worked almost every time.

When we first made human DNA libraries from people with normal globin genes, and from our first thalassemia patients, it was again heaven for the lab rat in me. I felt like I was visiting places no one had ever been before. I was an explorer reaching my own new North Pole or new moon. I was awed, excited, satisfied. When we had a new clone, it was like having a new baby.

A short time after we visited Pasadena, Maniatis and his team had cloned and sequenced all of the fragments of the normal human β globin locus, in part using our restriction data to organize their clones. Soon after, we could use Maniatis' data to obtain specific radioactively-labelled pieces of cloned DNA and use them, instead of our β globin cDNA, as our probes to identify more specific regions of the δ and β globin genes in our libraries. This made the analysis of the clones easier and more specific than using our β globin cDNA.

With the introduction of gene cloning, our studies of human β globin gene organization by restriction mapping became much less important. In less than two years (1978–80), restriction mapping as the way to analyze the detailed structure of DNA was history. With cloning, large amounts of pure DNA for any gene was in hand directly available to be sequenced and further studied.

Our novel restriction maps of the human β globin gene locus had largely served their purpose. Such is life in science. New technology means better ways of answering biological questions. We had had the spotlight for a few months, but the single base mutations in Cooley's anemia could be found consistently only by cloning, and it became the technique of choice.

When I came home from California and our visit with Maniatis in 1978, we set up gene cloning in the lab as soon as possible. Except for the hybridization of globin cDNA to DNA on filters, the steps in the cloning procedure were largely unfamiliar to us. We had never used bacterial cultures; bacteriophage; screening of bacteria; or replica plating in the way we needed to do so now. We broke down the experimental procedures into their individual new steps and went about perfecting each step.

We quickly made partial EcoRI human DNA libraries of our own from normal human DNA samples. Our normal human EcoRI libraries were not only useful to us as an entrée into cloning and sequencing new thalassemia genes, but we also shared them with others, as Maniatis had shared his library with us. Over the next several years, we cloned and sequenced several genes from patients with Cooley's anemia and identified as well as characterized the specific mutations in these genes, the ultimate steps in understanding β thalassemia.

I was fortunate at just that time, during the lab's cloning days, to have a terrific group of co-workers. Checco Ramirez used our new human gene libraries quickly to clone and sequence the human γ globin genes for the first time. Greg was still there. And there were several graduate students including Maryann Donovan-Peluso, Kathy Kosche, Sally Spence, and Carl Miller. Postdocs working on different cloning projects included Michael Baird, Norma Lerner, David Leibowitz, Cathy Driscoll, Kass Young, Debbie Rund, Alexander Burns, Carl Dobkin, and Bob Pergolizzi. Each of the graduate students and postdocs had a first authored paper in a peer-reviewed journal before they left the lab, and some much more.

Intervening sequences: a whole new biology

In the late 1970s, a new part of genes, so-called intervening sequences, was discovered. The new fact was that the globin genes, like almost all other mammalian genes, have non-coding nucleotide sequences that interrupt (or intervene) in the midst of the globin coding triplet sequences that specify the sequence of amino acids in the protein. The nucleotides that contain the triplet code for amino acids are called exons; the non-coding DNA sequences are called intervening sequences (IVS) or introns (Fig. 3, Appendix).

This discovery was made around the same time that we were using restriction analysis and gene cloning to study human globin gene organization and the gene defects in patients with Cooley's anemia. These intervening sequences provided a clearer understanding of how mammalian genes, including human globin genes, work, and β thalassemia turned out to be the poster child for demonstrating the biologic importance of these IVS.

The details of how these intervening sequences act in globin metabolism is not completely clear even to this day. There is evidence that these sequences interact with specific proteins in the nucleus of cells to perhaps protect and guide unedited immature longer mRNA transcripts from the nucleus to the cytoplasm of cells. Editing is then required to produce the mature useful mRNA from these longer transcripts for protein translation (Fig. 3, Appendix).

Intervening sequences were first identified in RNA. Early on, it was shown by restriction analyses and cloning of genes, including the human globin genes, that these intervening sequences are encoded in the DNA, interrupting the coding DNA triplets. These sequences are then transcribed into RNA, as usual, as part of an RNA precursor.

However, by the time globin mRNA moves to the cytoplasm of cells and becomes associated with ribosomes for globin synthesis (or translation), the introns must be removed or spliced out of the RNA transcripts (Fig. 3, Appendix). This RNA processing is necessary to ensure that the coding exon triplets (codons) are in a row, uninterrupted, to specify the proper amino acid sequence. If introns remained in the globin mRNA, they would interrupt the coding triplets and destroy the proper addition of amino acids.

What was also established early on as well, and by the work on thalassemia mutations, is that these IVS are required for normal globin protein to be made. Globin cDNA without IVS cannot make globin. IVS are required, but the reasons why this is true still remain obscure.

Human β globin protein has 146 amino acids. There are two introns in the human β gene, as in all human globin genes: a first intervening sequence (intron 1 or IVS1), and a second, intron 2, or IVS2. IVS1 is located in normal human β globin between the nucleotides coding for amino acids (codons) 30 and 31, while IVS2 is located between codons 104 and 105 of the β globin chain. I may not remember my neighbor's name, but I still remember the numbers of those codons involved in globin splicing.

Although the question of why IVS are present in genes persists, the process by which they are removed (or spliced out) is well understood. The 5' (left end) and 3' (right end) of each IVS specify the information necessary for the proper removal of IVS (splicing) to occur. The nucleotide sequences here, called consensus splice sites, are nearly the same in every globin gene (and in its corresponding globin mRNA precursor transcript).

Specific enzymatic processes, using unique proteins, splice out the IVS precisely from the precursor RNA with these consensus splice sites as their guide, to produce the mature translatable mRNA (Fig. 3, Appendix). I also still remember from the early 1980s that AGGT is the consensus splice sequence for the 5' end of human β globin IVS2. We and others discovered that a single mutation in any one of these four bases leads to abnormal splicing and an abnormal β globin mRNA, and results in β thalassemia. We and others showed subsequently that many other single base mutations in IVS result in either β zero or β plus thalassemia.

So whatever the roles of IVS in the nucleus, and whatever processing events occur that lead to the generation of mature globin mRNA, it is clear that the proper removal of these IVS is required for normal human β globin to be produced.

We were fortunate that IVS were discovered just as we started cloning normal globin genes and those in thalassemia, otherwise their presence would have confused us as we tried to decipher the meaning of the DNA sequences we cloned. Until their discovery, we had assumed that the sequence of the human β globin gene was simply the same as that of the β globin cDNA sequence, because we had made the cDNA from mature globin mRNA. Until the discovery of IVS, we felt that we knew what DNA sequences were required to transcribe β globin DNA into RNA and make globin protein. As it turns out, IVS are indeed a critical additional part of this process.

Sally Spence's gene

One of our problems in the early days of cloning was that we had no way of knowing whether a new cloning project would lead to the cloning of a new thalassemia gene with a new mutation, or whether we would find a gene with a mutation that had been cloned and sequenced before. Repetition is not the goal of research. I never liked repeating anybody else's work, or publishing the same findings that others reported, even though, when that occurs, it confirms the validity of the initial original results. My goal was always to discover something new. And yet there was no way up to that time to ensure that our cloning and analysis of genes in a new thalassemia patient would not lead to the rediscovery of an already described mutation. A frustrating situation for us.

A single base change at the nucleotide triplet coding for the 39th amino acid of human β globin, codon 39, had first been shown by Dr. Yuet W. Kan's group to cause β zero thalassemia. A single change in a DNA base at codon 39 led to a triplet that did not code for any amino acid. Instead, it led to the termination of the growing amino acid chain on the ribosomes. The codon 39 mutation is very common in certain thalassemia populations and is the most prominent mutation in Sardinia. Codon 39 mutations, unhappily for us, showed up repeatedly in our early cloning of DNA from thalassemia patients. In collaboration with Kan, we published a couple of them early on.

In 1980, we were finally lucky enough to find one unique thalassemia gene, and we spent the next several years studying it. It revealed a unique mechanism by which human β globin genes work to make normal amounts of human β globin, and provided a new model of how single base mutations within IVS can regulate splicing.

The cloning and sequencing of this gene in our laboratory was almost entirely the work of Sally Spence, a graduate student at that time. Sally is a tall woman, about 5'11", and usually smiling, but tough as nails, intense, self-critical, and an extremely hard worker. Sally was determined to find something new for her thesis and she finally did. She was lucky enough to clone a most interesting gene.

In these studies, she worked with Dr. Gekee Sim who, at the time, was a postdoctoral fellow with Dr. Richard Axel at Columbia. Gekee, who was brilliant, had worked with Maniatis earlier, and was the most knowledgeable person I knew with regard to the cloning technology. But Gekee's goal was absolute perfection. I always say, you can't be too compulsive in science, but compulsivity has its limits, and Gekee tested those limits. I could see Sally's mix of determination, admiration and frustration as Gekee peppered her with questions as they worked together closely day after day.

"Are you sure you grew the bacteria right? Do you really think that spot is the same on those two filters? Where are the results? Show me the plates? Are you sure you didn't forget to wash the filters?"

There was Sally, 5'11", big and tough, and Gekee, much smaller and even tougher, barking out orders and questions: two interacting scientists working for the same goal and collaborating successfully. But I would not have been surprised to come into the lab one day and see Gekee lying bleeding on the lab floor or otherwise physically maimed for the way she harassed Sally in the lab. I was happy no bones were broken, but I always felt that feelings were hurt and that there was destructive tension in their relationship.

I was on the scene often enough at that time to see it all first-hand. Sally was in tears at times, but never complained. Even though I was the boss, I felt Gekee's wrath as well. But I could tolerate almost anything in the lab if it contributed to the lab's success, and Gekee helped accomplish that. In addition, outside of the lab, Gekee was one of the warmest and nicest people I knew.

Sally's thalassemia gene turned out to be quite complicated and confusing. It had five base changes in its DNA sequence from the normal β globin gene, all within the second intervening sequence of the patient's β globin gene; many more changes than we knew how to interpret. She presented her data at a meeting, and, shortly thereafter, Dr. Stuart Orkin from Harvard, one of the stars in the human globin field who discovered a protein called GATA-1 (a critical transcription factor in red cell development), saved us from further frustration and confusion.

In those days, well before e-mail, he sent me a hand-written note saying that he and his collaborators had discovered that some of the DNA base changes in Sally's gene were also found as normal variations in many normal human β globin genes from different individuals. These single nucleotide changes that are variations from "normal", but which have no pathological consequences are called polymorphisms (or neutral polymorphisms). In different populations, there are different polymorphisms. Human DNA polymorphisms occur at a high rate: about 1 in 100 of our DNA bases are polymorphic with no known adverse consequences.

Polymorphisms in human DNA can be used to trace different human populations as they diverge. Orkin and Dr. Haig Kazazian at Johns Hopkins and their co-workers, had found several different combinations of neutral polymorphisms that identified different ethnic groups with different β thalassemia mutations. The variations in these polymorphisms were used by these and other investigators to limit the chances that a new thalassemia gene being cloned would be the same as a previously analyzed one. If there were different polymorphisms at the β globin locus in a patient from which a new thalassemia gene was being cloned (as compared to that of patients with genes previously studied), it was unlikely that the new gene would have the same single base mutation as the previous genes; although this assumption was not fool-proof.

Polymorphisms have also been used extensively in forensic analyses, to determine the source of the DNA of a fetus or of any individual in paternity testing and criminal cases, like rape.

With Orkin's information pinpointing the most likely polymorphisms in Sally's gene, we could finally focus on one nucleotide change, a mutation at position 705 of β IVS2, as the probable cause of the thalassemia. But how did it affect β globin production? That was a question we had to answer before we could be sure of the significance of the 705 mutation in Sally's gene. After many experiments over several years, Dr. Carl Dobkin, a postdoc in the lab, eventually proved conclusively that the single base mutation at position 705 of IVS2 prevented normal splicing, and was responsible for the lack of human β globin production by this gene.

Details, details

The detailed biochemical analysis of Sally's gene and how it led to thalassemia is difficult for even the knowledgeable scientific reader to understand. At the very least, it requires the reader's interest in the details. Of course, I think it's a terrific biological and biochemical lesson in how genes work and how this single base mutation can cause thalassemia, but you may not be as excited by the details of the story. If that's true, please feel free to move to the next section because I won't spare the details as I tell the whole story.

We suspected that the single base change at position 705 of IVS2 was the culprit in Sally's gene leading to thalassemia early on, largely by inductive reasoning alone (a dangerous scientific method in my experience). In this speculation, we thought that the single base change at 705, from ATGT to AGGT, resulted in the creation of a new normal consensus 5′ splice site in this gene. Instead of all of IVS2 being removed in one step, this new 5′ splice site created by the mutation led to a new pattern of splicing (Fig. 4, Appendix). A new splice was made from the 705 mutation to the normal 3′ end of IVS2. This single base change led to bad editing of the mRNA transcript, analagous to an editor crossing out only some of the wrong words in a critical sentence when the whole sentence had to be deleted. Instead of removing all of IVS2, the editor of the RNA (in this case due to the single base change) crossed out only part of IVS2 and left the rest in.

The β globin IVS2 is 920 bases long. Normally, all 920 would be spliced out in a single step. However, because a new AGGT 5′ splice site was created by the 705 mutation, this favored only partial splicing: from the new abnormal splice at 705 to the normal 3′ end of β IVS2 in this gene at 920.

But that wasn't all that went wrong. We recognized that a DNA sequence normally present in IVS2, at position 580, could serve as an alternative 3′ splicing site. We thought that what happened in Sally's gene was that this 580 sequence became a 3′ splice partner to the normal 5′ splice site, perhaps because of the occupation and unavailability of the normal 3′ end of IVS2 (tied up in splicing by the 705 mutation). Thus, 580 was a default 3′ splice site for the 5′ end of IVS2, because its normal partner, 920 nucleotides away, was now usurped by the newly-created 5′ splice, closer, at 705 (Fig. 4, Appendix).

So in summary, we thought that the 705 mutation somehow intercepted or prevented normal splicing of IVS2 by leading to these two abnormal splices, one from the normal 5′ site of IVS2 to 580, the other, 705 to 3′ end of IVS2, to produce an abnormal globin mRNA and absent β globin and β zero thalassemia (Fig. 4, Appendix).

If our reasoning was correct, when the two abnormal splices involving 580 and 705 occurred, a piece of IVS2 from 580 to 705 should remain in the mature β globin mRNA, making it elongated, abnormal and untranslatable. This explanation was quite unique at the time, and largely theoretical, at first.

Dr. Carl Dobkin eventually proved experimentally that all of our theoretical assumptions of how Sally's gene worked were true! Carl was an excellent scientist and a wonderful colleague in the lab. He was a poker-faced dead-panned guy with a self-effacing demeanor and a mischievous smile on his face most of the time, except when he was really annoyed, as when anyone else in the lab borrowed his reagents without telling him. RNA, especially mRNA, was easily degraded and he didn't want anyone contaminating his reagents. Carl was a true lab rat.

Carl always had a chuckle at the ready and a droll sense of humor, and he always knew the next experiment to do without much direction. We got along well together, as well as I ever have with anyone in the lab. He was as single-minded as a tiger and would not settle for anything less than absolute proof as he pursued the precise explanation for how Sally's gene worked. I was as proud of his original and unique work on this gene and the new mechanism we discovered as I am of any of the science that I have done.

Carl established the details of the alternative splices in Sally's gene by using new technology referred to by others, (not by me), as "reverse genetics." Webster's dictionary says: "In classic genetics, the traditional approach (in the field) was to find a gene product and then try to identify the gene itself. In [some] molecular genetics, the reverse has been done by identifying genes purely on the basis of their position (structure) in the genome with no knowledge whatsoever of the gene product. This revolutionary approach is called "reverse genetics." I never liked the phrase.

In this technology, as applied to globin, one took the DNA of a cloned normal or thalassemia gene of interest, together with its regulatory sequences and other DNA sequences that made it function, ligated them into a plasmid (a piece of DNA), and added the plasmid to cells in which the globin gene could be expressed as mRNA. Hela cells, cultured human fibroblasts (non-globin-producing cells) were commonly used to express the gene's function. These cells could be grown to large amounts, and the globin mRNA produced by the cloned gene could then be analyzed for its size, using radioactive globin cDNA probes.

What Carl found was that in Hela cells, while addition of normal human β globin gene DNA produced only normal-sized globin mRNA, Sally's cloned gene DNA produced only a larger-sized mRNA. This larger-sized piece contained the RNA sequences between IVS2 positions 580 and 705, as we had predicted, and was untranslatable. It could not make β globin protein.

Carl then proved that the abnormal splicing of Sally's thalassemia gene could be corrected so that normal splicing could be restored. He did this by simply mutating only the 580 base of Sally's gene (present in the normal gene) so that it lost its ability to serve as a cryptic 3' splice site. What he found was that only normal globin mRNA was produced when he put this new 580 altered gene into Hela cells. With no cryptic (or alternative) 580 site available, the normal 5' splice site found its normal 3' end all the time; even though Sally's gene still had the 705 mutation! Carl had cured the thalassemia defect by making a base change not at the site of the mutation (705), but in a normally present cryptic splice site (580). An elegant molecular tale, if I say so myself.

Sally's gene was the first β globin gene found to cause thalassemia that had a mutation within human β globin IVS2, but still had normal 5′ and 3′ consensus sequences at its ends; and yet it led to abnormal splicing, no normal β globin mRNA, no β globin, and Cooley's anemia.

Other cloning tales

During the 1980s, our lab cloned and sequenced several other novel β thalassemia genes with defects in β IVS 2 sequences, as well. One of these had a mutation close to the 3′ splice junction that led to thalassemia. Another was a gene with a single base change at position 5 of β IVS2. None were quite as exciting as Sally's gene, but all were interesting and new.

Maryann Donovan-Peluso also used the Hela cell system to good advantage, in analyzing human β globin gene function in human K562 cells, a cell line that expresses human γ globin genes, but not human β globin genes. Maryann was one of my best friends in the lab. She worked for me first as a technician, then as a graduate student, and finally as a postdoc. We had lots of lunches together, and each time she insisted on hot French fries. "Hot, please." She also was really thin. She taught me to "eat only what you really enjoy. Nothing else."

One of her projects was to determine whether the absence of β globin expression in K562 cells was due to the presence of an abnormal β globin gene, or to the specific repression of the expression of a structurally normal human β globin gene in these cells. When Maryann cloned the human β globin gene from K562 cells and expressed it in Hela cells, she found that it led to normal β globin mRNA function, indicating that it was a normal β globin gene that was somehow silenced by unknown mechanisms, still unknown today.

By the late 1980s, we had also used globin gene cloning and gene expression in Hela cells to analyze the role of different regions of the normal human β globin locus genes. We mixed and matched pieces of the normal human γ, β and δ globin genes, and showed that some pieces were functionally interchangeable and others were not.

Dr. Suzanne LaFlamme, a postdoc in the lab, found that when a large restriction fragment containing most of IVS2 of the δ globin gene was exchanged for the equivalent fragment from β IVS2, in an otherwise intact human β globin gene, β globin mRNA expression was markedly reduced. Also using Hela cells, Maryann showed that "enhancer" sequences, present in the region 3' of the β globin gene, when added to the human γ globin gene, worked as well there in increasing γ globin gene expression as when attached to the human β globin gene.

All of these experiments provided new insights into the multiple levels of regulation controlling the expression of the human globin genes. They were keys to understanding thalassemia, and provided new perspectives that might be used for future globin gene therapy.

These detailed studies of gene structure and function in Cooley's anemia by our lab and several others showed that many different single base mutations could result in the same pathological state, in our case, anemia due to decreased or absent human β globin, by many different mechanisms. Studies of sickle cell disease had previously shown that a single base change in the DNA sequence of a human β globin gene could lead to a structurally abnormal globin protein and cause a human disease. With the sickle cell mutation, there was a change in the quality of the protein. But studies of thalassemia were the first to demonstrate that there were many different single base mutations, every one of which could lead to changes in the quantity of a normal gene product, here, normal human β globin, and every one of which could result in the same human disease, in this case Cooley's anemia.

These thalassemia studies identified critical single base changes in: (1) regulatory sequences upstream of the β globin genes that result in decreased or no β mRNA; (2) coding triplet sequences that prevent abnormal globin mRNA from being translated into globin protein, the production of so called "termination" or "nonsense" codons (like codon 39 mutations); (3) non-coding β globin gene sequences (intervening sequences) that prevent the proper processing of β globin mRNA (like Sally's gene); and, (4) the polyA tail of the globin mRNA that prevents normal globin protein production. All of these different kinds of single base changes result in decreased or absent β globin. When two such mutated genes are inherited in one individual, the result is Cooley's anemia.

The specific kinds of molecular defects first discovered in the thalassemias led to the subsequent discovery of the same specific genetic mechanisms causing many other genetic and acquired diseases. The molecular defects discovered in the thalassemias, however, never received the same amount of interest in the wider scientific community as when these same mechanisms were discovered later in other more prevalent disorders like heart disease or cancer.

Why so many mutations

Can we explain why so many different genetic mutations that cause decreased β globin and thalassemia exist in different populations around the world? Is there genetic pressure for red cells with decreased or abnormal β globin to survive? Is there an advantage for individuals to inherit one abnormal β globin gene, as occurs in people with thalassemia trait (and sickle cell trait as well)? The answer to all these questions under certain circumstances is "Yes."

The presence of one abnormal sickle or thalassemia gene (sickle or thalassemia trait) appears to "protect" people from dying from falciparum malaria, a severe and potentially fatal form of malaria. People with thalassemia trait, like those with sickle cell trait, have a lower mortality from severe malaria than those of us with two normal β globin genes.

How does this happen? It turns out that the malarial parasite proliferates well in normal red blood cells, and causes them to burst, thus spilling large amounts of the parasites into the blood stream. This proliferation spreads the parasites through out the body, to infect more cells, and causes severe disease. By contrast, the affected red cells in people with sickle and thalassemia trait are destroyed more quickly than normal red blood cells, so the malarial parasites do not have as much time to multiply as they do in normal cells; the trait cells are poorer hosts for the parasites, and, therefore, there is less severe disease and less mortality.

Heterozygosity, the presence of one abnormal gene for sickle or β thalassemia, is an example of a so-called "balanced polymorphism." This phrase refers to the utility of having mutant genes that favor survival under certain circumstances, like those in malarial regions. That is why so many different mutant β globin genes arise and are selected to survive in the context of malaria.

The hypothesis is: People with two genes for sickle cell or β thalassemia have severely shortened life spans, especially before modern treatment. Patients with these conditions would presumably disappear from human populations relatively quickly. However, individuals with one mutated gene for β thalassemia trait or sickle cell trait survive more often in the areas of severe malaria than those with two normal β genes. This phenomenon counterbalances the increased mortality of patients with two abnormal genes (homozygotes), and leads to the persistence of the mutant heterozygous genes in the population.

* * *

In summary, in the late 1970s and 1980s, we and others utilized a remarkable new method, gene cloning, to isolate and characterize normal human β globin genes and those of patients with Cooley's anemia. These studies resulted in the discovery of many new single DNA base mutations in the human β globin gene in these patients that are the root cause of the disease, and that lead to decreased or abnormal β globin mRNA, decreased or absent human β globin, and anemia.

Many of these mutations have been shown to be expressed through unique molecular mechanisms that cause Cooley's anemia. Cooley's anemia served as the model for defining different types of molecular alterations in gene structure that lead to abnormal gene function in many other human diseases.

Chapter 12

You Can't Go Home Again

As discussed earlier, understanding the regulation of human fetal hemoglobin is central to the problem of trying to cure Cooley's anemia. As in the rest of us, fetal hemoglobin is produced normally in patients with the disease during fetal life. However, as also is the case with the rest of us, the production of γ globin and fetal hemoglobin (HbF) is essentially turned off in late fetal life and there is a switch to the production of adult hemoglobin (HbA), which contains human β globin. In Cooley's anemia, the defects in β globin production during this switch cause the disease.

Hemoglobin switching is a complex event. Several different steps, many presumably still unknown, are probably necessary to initiate γ globin synthesis in red cells in fetal life, and then to turn it off in adult red cells. In addition, different steps are also required to maximally activate β globin production in adult red cells.

These steps involve interactions between a special control region upstream (to the left) of the structural genes (embryonic, γ, δ and β genes) at the β globin locus, and the DNA in the vicinity of the genes themselves (Fig. 1, Appendix). They also involve the actions of different proteins that regulate the specific expression of each of these genes at the fetal and adult developmental stages. These complex interactions between genes and proteins control globin switching.

In this chapter, I will describe one such group of regulatory proteins, a complex of proteins we call PYR complex, that we believe has a significant role in human hemoglobin switching and in the regulation of human fetal hemoglobin and that has been a focus of my laboratory research for the past two decades. PYR complex binds to sequences upstream of the human δ globin gene in the γ-δ globin intergenic region that we have previously shown to be important in regulating human fetal hemoglobin.

PYR complex appears only in adult blood cells and may function in the switching process by repressing human γ globin gene expression and activating human β globin gene expression. If switching could be reversed and fetal hemoglobin could be reactivated to high levels in adult red blood cells, this reactivation would compensate for the lack of β globin in Cooley's anemia and could eventually cure the disease.

*   *   *

Controlling globin switching

As I've mentioned before, fetal hemoglobin is normally used to supply oxygen to the fetus. Fetal hemoglobin (HbF, $\gamma_2\beta_2$) contains two γ globin protein chains.

There is a dramatic switch in late fetal life, from the production of γ chains to β chains, from HbF to HbA. Why do we need γ globin and fetal hemoglobin? It is known that HbF has a higher affinity for oxygen than HbA. Its γ globin fingers hold the oxygen in hemoglobin tighter than the β fingers of HbA. That means that as blood runs through the placenta, the HbF-containing red cells of the fetus can grab oxygen preferentially from the mother's HbA-containing blood. This favors oxygenation of the fetus, and the persistence of HbF in humans strongly suggests that γ globin production evolved to promote preferential survival of the fetus.

The presence of fetal hemoglobin also explains why patients with β thalassemia survive fetal life and are born healthy, despite having little or no β globin production. During fetal life, as with normal fetuses, fetuses with thalassemia don't need β globin; they use their normal γ globin genes to make normal amounts of HbF.

The major problem with fetal hemoglobin regulation is that once its production is shut down, it can almost never be reactivated to any significant extent. Only small amounts of fetal hemoglobin continue to be made in adult cells. For unexplained reasons, we can't go home again; we can't make much more fetal hemoglobin when we need it, even though that would be the cure for β thalassemia.

The importance of this problem and its intuitively obvious solution, treatment leading to the full reactivation of fetal hemoglobin production, while recognized over 40 years ago, has not resulted in any clinically meaningful solutions to date.

However, many insights into this problem have been gained by both experiments of nature and by interventions with drugs that attempt to reactivate fetal hemoglobin. Amazingly, as I have mentioned earlier, there are a few people who produce no β globin at all, who make only γ globin and HbF in adult life, and have no anemia! They never switch to β globin production, in contrast to the rest of us, including patients with β thalassemia.

Why and how does this happen? The short answer is that we still don't know. The longer answer is partly below, and is based on research some of us have done for years involving the molecular events in human hemoglobin switching which we are slowly beginning to understand.

As I have indicated before, the genes in DNA lead to mRNAs which lead to proteins such as globin. The human γ, δ and β globin genes are arranged on chromosome 11, with the embryonic and fetal genes at one end, and the adult β globin genes at the other (Fig. 1, Appendix). The control of the expression of each of these genes is complex. Except for the sperm and egg cells, every cell in the body contains the same DNA and genetic material. However, we only express certain specific genes in specific cells as their functions are required. As I've mentioned before, we don't want hemoglobin expressed in our eyeballs any more than we want or need eye proteins expressed in our red blood cells. Also, in terms of the biological work to be done by a particular cell, there is an economy to be gained by having only those proteins we need at any given time expressed in specific cells.

In typical cells in an organ, like red blood cells in the bone marrow, most genes are normally repressed and do not express most of the proteins they can potentially produce. In addition, different genes are repressed or expressed at different stages of development and maturation in different cells. In the red blood cells in the fetus, the γ globin genes are expressed, and the β globin genes are repressed. Then around birth, the switch occurs, the γ globin genes are repressed, and the β globin genes are activated.

The answers to why and how this selective repression and activation occur is, as usual, in the details of these processes. What events turn on the human γ globin genes in the first place in the red blood cells of the fetus? And what events then repress these genes and activate the β globin genes around birth?

The human β globin locus control region (βLCR), discovered by Dr. Frank Grosveld in 1987, is a major player in both hemoglobin expression in red cells in general, and in human fetal to adult globin switching. The βLCR is a DNA sequence far upstream (5′, to the left) of the human γ and β globin genes whose action dramatically increases the expression of the specific globin gene with which it associates (Figs. 1, 5, Appendix). When the βLCR and its associated specific proteins become physically associated with proteins near either the structural γ or β globin genes, the globin protein output of that gene is amplified 100 to 1000 fold. The βLCR increases the expression of globin genes in a manner analogous to the volume control mechanism on a radio, increasing the loudness of sounds we hear.

The βLCR is brought into proximity of either the human γ or β globin genes downstream (to the right) through interactions with regulatory proteins, so-called "transcription factors" (Fig. 5, Appendix). These factors are specific regulatory proteins expressed in erythroid cells, with names like GATA-1, NF-E2 and EKLF.

These interactions between proteins and DNA leading to the proximity of the βLCR to individual globin genes occur by "looping of DNA" in the structure of chromatin (Fig. 5, Appendix). The βLCR is a long way away physically, and the way it loops around to become in contact with either the γ or β globin genes, must be programmed by specific molecular events.

It has recently been shown experimentally by Grosveld and others that during fetal life, the β globin LCR is physically brought in contact with the human γ globin genes by looping, and increased fetal hemoglobin production is the result. Then, in adult cells, there is the switch and the β globin LCR comes in closer proximity with the human β globin gene instead of the γ globin gene and adult β globin and HbA result (Fig. 5, Appendix).

It is known that, in general, the group of specific proteins called transcription factors, that bind to genes in a complex structure called chromatin, control specific gene expression. Chromatin is the chromosomal environment of genes and contains complex three-dimensional arrangements of proteins and DNA.

The red blood cell specific transcription factors like GATA-1, EKLF, and NF-E2, are required for normal red blood cell development and optimal hemoglobin production. They are involved in the looping interactions between the βLCR and the γ or β globin genes. However, these red cell transcription factors are active in both fetal and adult red cells, and cannot, by themselves, be responsible for human globin switching. Similarly, erythropoietin, the major hormonal growth factor for red cells, is active in both fetal and adult red cells.

There must be stage-specific protein factors or stage-specific modifications of existing proteins that control human globin switching. From our research, we think that the adult stage-specific protein, Ikaros, is one of these proteins. Ikaros is in a complex that we discovered called PYR complex, to be discussed below (Fig. 5, Appendix). Switching clearly does not consist of a single step; it is not a single action switch, like a light bulb being turned on or off; and there must be other stage-specific factors and complexes. The details of the entire process are still relatively obscure.

Other molecular events can influence gene expression and repression, including the process of human globin switching. Small changes in histone proteins, which coat all DNA in chromatin, can modify specific gene expression significantly. An explosion of new findings in recent years indicates that so-called epigenetic changes such as the addition of tiny chemical subunits, including methyl groups or acetyl groups, at precise positions on histones associated with specific genes can dramatically alter the expression or repression of these genes. Histones alter the configuration of the protein-DNA structures in chromatin, and thereby change the availability of gene sequences to their regulatory proteins.

The addition of acetyl groups to histones, histone acetylation, is associated with increased gene expression at sites where this acetylation occurs. Conversely, histone deacetylation is associated with specific gene repression at the sites where deacetylation takes place.

Also, the addition or loss of non-histone proteins, the transcription factors, interacting with DNA and/or histones, sometimes as part of larger protein complexes, can upregulate or downregulate specific gene expression. These complexes bind to DNA sequences within or between specific genes in chromatin, and can change specific gene expression.

The proteins comprising the basal transcriptional machinery required to transcribe DNA into RNA, including RNA polymerase, are another critical component in gene expression. This machinery is used by all genes and, with DNA and histone and non-histone proteins, is part of large complex structures specific to each particular gene, called enhanceosomes; these three-dimensional structures control gene expression.

Additionally, relatively newly described components, small nuclear RNAs transcribed from DNA, called interfering RNAs (siRNAs) or micro RNAs (miRNAs), are also active in regulating specific gene expression.

In the end, specific genes in different cells are activated and expressed when the structural relationship between regulatory transcription proteins and the histones and DNA in chromatin is right; and expression of specific genes is repressed when the configuration of DNA and these proteins in chromatin is changed. In different cells at different stages of their development, the specific structural configuration of DNA and proteins in the nucleus leads to the specific expression of certain genes and the specific repression of others.

Over the past 20 years, a few stage-specific protein complexes have been reported that can help explain human globin switching. Two of these, the discoveries of other laboratories, affect the fetal globin stage alone: one such protein complex activates γ globin gene transcription in fetal cells, and the other complex reported represses γ globin gene expression in adult cells, presumably by modifying the interactions between the βLCR and the human γ globin gene.

We have described another protein complex that we call "PYR complex" that we believe is a potentially true "switch complex." This complex is present only in adult blood cells. We call it a switch complex because we think it acts to both specifically repress human γ globin expression, and, at the same time, activate human β globin gene expression in these adult cells. PYR complex is presumably just one of the protein complexes working to maximize switching.

The rest of this chapter describes our work over the past 15 years on PYR complex and its interactions with DNA at the human β globin locus, and how these events affect human γ to β globin switching.

As usual, our experiments were driven by the development of new technologies, this time in the late 1980s. Along with DNA cloning and sequencing, new ways of assessing the interactions of proteins with DNA in mammalian cells became available. As mentioned earlier, chromatin is the chromosomal environment of genes, and contains complex arrangements of proteins and DNA. New methods were described to precisely define the interactions between specific proteins and specific nucleotide sequences of DNA.

Finding something new

Research on chicken red blood cell development by Dr. Doug Engel, in the 1980s, was extremely intriguing to me as we began to search for proteins that might be involved in human globin switching. Engel showed that there is a specific "regulatory" DNA sequence which controls chicken globin switching between the chicken embryonic globin genes (chicks have no fetal genes although the embryonic genes substitute for them) and the adult chicken globin genes. Specific protein transcription factors interacting with this regulatory sequence appeared to control the "switch" from the chicken embryonic to adult globin.

By this time, the late '80s, the complete sequence of nucleotides at the human β globin locus, from far upstream (to the left) of the embryonic and fetal human globin genes, to far right and downstream of the adult human β globin gene, had been obtained by gene cloning and sequencing (Fig. 1, Appendix). About the same time, computers became easier to use, at least for me, to analyze the potential biological significance of these sequences.

As I indicated earlier, my previous work had suggested that the human γ-δ intergenic region might control human globin switching, at least in part. In the 1970s, we had shown that in patients with δ-β thalassemia and HPFH in whom this intergenic region is either partially or completely deleted, γ globin and HbF are markedly increased.

Driven by these new developments and our interest in human globin switching, Dr. David O'Neill, a postdoc in my lab, and I asked the question: Are there any DNA sequences at the human β globin locus that look similar to the globin switch sequence in the chicken that Doug Engel had described? There were a few candidates identified by the computer, but the most interesting one was in a region between the human γ (fetal) and β (adult) globin genes, a region upstream of the human δ globin gene, the γ-δ intergenic region (Fig. 1, Appendix).

David used a new technology available at the time to investigate the potential biological role of this intergenic region upstream of the human δ gene. This technique is called "gel retardation analysis" or gel shifts, and it measures the interaction between specific DNA sequences and the nuclear proteins binding to them. We were looking for red cell nuclear proteins that controlled human globin switching.

Experimentally, gel retardation analysis is fairly straightforward: (1) pieces of DNA for study are made radioactive; (2) the radioactive DNA pieces are mixed with proteins extracted from the nuclei of the relevant cells (nuclear extracts); and, (3) the free DNA is then separated from DNA bound to the desired proteins by chromatography on polyacrylamide gels.

The DNA pieces used for study are called "oligonucleotides," or oligos. They are 20 to 250 nucleotides long, can be chemically synthesized, and were (and are) available commercially. We first evaluated DNA sequences from the γ-δ intergenic region homologous to the Engel chick switch sequences. We made these γ-δ oligos radioactive so we could detect even a small amount of protein-DNA complexes, wherever the complexes migrated on the gels. A small amount of colored dye is also added to the gel samples in order to see how far down the gel even a small molecule like the dye migrates and to insure that our radioactive oligo does not run off the bottom of the gel.

When an electrical field is applied, the radioactive DNA alone, of small size, runs down the gel quickly, close to where the dye goes.

But if bound to specific protein (s), the movement of the radioactive DNA in the gel is slowed by its binding to nuclear proteins; it is "retarded" or "shifted": ergo, the name, DNA gel shifts. We identify the retarded bands by measuring their radioactivity using autoradiography as usual, i.e., by placing the gel against an x-ray film and having the radioactivity on the gel, wherever present, darken the film after this exposure. These were our "gel shifts."

These were fun experiments in a new unexplored world, the kind I like best. Are there specific nuclear proteins bound to our γ-δ intergenic sequences? If so, what are these mysterious proteins? How do they work? Are they specific for blood cells or not?

What we found in these first experiments was that the red blood-forming cells we used, indeed, contained specific nuclear proteins that interacted specifically with one of our γ-δ intergenic region oligonucleotides, not precisely at the same region as Engel's chick switch sequence, but quite close to it. As controls for the specificity of our DNA-protein interactions, we used oligos from different γ-δ intergenic regions and others, far away and unrelated to γ-δ intergenic sequences. We found no such interactions, no protein binding between our nucleated red cell extracts and these unrelated oligos.

The specific oligo that interacted with proteins best was composed of 200–250 bases that were almost exclusively "pyrimidine" bases in DNA, thymine (T) and cytosine (C), and not so called purines, adenine or guanine. The DNA in the sequence was almost all C's and T's,...CCCTTCCTTT...etc. For this reason, we called this pyrimidine-rich sequence, PYR sequence, and the proteins bound to it, PYR factors or PYR complex.

In these first experiments, we used extracts from mouse erythroleukemia cells (MELC), a cell line that is unique in that it produces large amounts of adult-type mouse hemoglobin, similar to HbA with its β globin in human adult red cells. MELC is essentially an adult β globin expressing cell line and produces little or no mouse embryonic hemoglobin.

The reason we used MELC was twofold: It was (and still is) difficult to obtain large amounts of purified human adult nucleated red blood cells from any other source that could provide the large amounts of purified nuclear extract required for protein isolation. Bone marrow, while most desirable, cannot be used because it is contaminated with white blood cells and other types of cells. In contrast, we could grow large enough amounts of MELC in tissue culture to purify and characterize the components of PYR complex, the regulatory proteins we were seeking.

After finding PYR complex in MELC, we asked whether this complex was also present in other types of cells. Nuclear extracts from human K562 cells, a human erythroid cell line that produces only embryonic and fetal human hemoglobins had little PYR complex as assessed by gel shift with the radioactive PYR sequence-containing probe.

Similarly, using nuclear extracts from non-hematopoietic cells, Hela cells (human-derived fibroblasts), we found no PYR complex. The only additional human cells that we studied that contained PYR complex were other adult-type blood cells, particularly human T-lymphocytes.

In another important experiment, we discovered that PYR complex was present in adult-type red blood cells of mice. Because the fetal liver in the mouse is almost exclusively composed of adult-type nucleated red blood cells, we were able to use this naturally purified source of adult red blood cells to show that these cells had high levels of PYR complex. This result demonstrated that real adult-type red cells, not just those in a cell line like MELC, contained PYR complex. By contrast, when we isolated nuclear extracts from mouse yolk sac cells, the equivalent of human fetal-embryonic red cells, we found little or no PYR complex.

Thus, we had discovered the first adult-type red blood cell complex, PYR complex, that was adult-stage specific and that specifically interacted with a sequence of intergenic DNA (PYR sequence) in a region of the human β globin locus that we suspected of being important in human fetal to adult globin switching.

We also showed in these early experiments that when PYR sequence was bound to PYR complex, their interaction changed the configuration of the DNA involved; the interaction of PYR complex with PYR sequence prevented the degradation of the DNA of PYR sequence by an enzyme called a nuclease, which normally degrades uncomplexed or naked DNA in chromatin. These data suggested that the interaction of PYR sequence with PYR complex in adult-type red blood cells alters the physical configuration of DNA between the human γ and β globin genes, and perhaps modifies the function of this DNA during human hemoglobin switching.

We published these discoveries in 1991, and hypothesized that PYR sequence and PYR complex had a role in human globin switching because their interaction: (1) was a specific interaction occurring only in adult-type blood cells; (2) represented an interaction involving a unique DNA sequence between the human γ and δ genes (in our suspected γ-δ switch region) that was likely to be involved in globin switching; and (3) changed the configuration of the DNA in the γ-δ switch region in adult-type cells.

We proposed in the 1991 paper that the interaction of PYR complex with PYR sequences in DNA might promote human globin switching by changing the configuration of chromatin at the human β globin locus, thereby repressing γ globin gene expression and perhaps, in addition, activating human β globin gene expression in adult-type red blood cells (Fig. 5, Appendix).

There were many unanswered questions resulting from this early work. What were the proteins in PYR complex? How did they work? What did they really do in terms of affecting globin switching in cells and animals? Also, although the DNA sequences we studied were from human cells, the extracts we used were from mouse cells. What relevance did mouse cell PYR complex have to human globin switching?

David O'Neill was another solid superb driven scientist I was fortunate enough to have as a colleague, and he was unswerving in his determination to find out whether PYR sequence and PYR complex were important in human globin switching. He stuck with the project from 1991 to 2004, and his diligence finally paid off. David was a careful and creative experimentalist and developed many techniques new to our lab to move us forward. Although he was a trained clinical pathologist who could have made much more money practicing his medical specialty much earlier, David was unquestionably in the ranks of the best and the brightest lab rats I have known. He loved the process of scientific discovery and was fascinated by the problem of human globin switching. He was there all day every day trying to figure out what to do next and doing it.

We had a wonderful group of graduate students and postdocs who assisted David along the way, but he was the main man. Some of the best of these students were in their first year in the Department of Genetics and Development at Columbia University, and only stayed in the lab for a few months. Others, like Rocio Lopez and Stuti Schoetz, spent years working on PYR. Dr. Michael Flamm and Katherine Bornschlegel, a technician, also contributed significantly to the PYR research.

The two major questions in 1991 were: (1) What is PYR complex, and (2) What is the biological role of PYR complex in interacting with PYR sequence? We tried to answer both questions at the same time by working on parallel research tracks.

It is amazing to me that I can summarize the 15 years of work David and the lab did on globin switching and PYR complex in one chapter. That's how science works. You never know how long it will take to get your answers. And you just have to persevere and push along with the best experiments you can do with the best available technology, and hope you get some answers and remain funded.

I have been lucky enough to have had continuous NIH support throughout my career, even up to the present time. You really can't do research on a high level, in areas such as those I have been working in, without certain critical ingredients: lots of money; a relatively big lab (five to ten people); really good people at the bench; more than one NIH grant; and publishable results in a year or two. These are tough requirements.

It took eight more years, 1991 to 1999, before David O'Neill would answer some of the questions his earlier discoveries posed, and before we learned anything more that was publishable; and it took even longer than that to show the biologic significance of our findings.

What is this stuff anyway?

From 1992 to 1997, David isolated PYR complex from MELC. We could grow liters of MELC, while we could not grow enough cells to do the same experiments with primary mouse or human blood cells. The assay during this purification of PYR complex was indirect. We followed the presence of PYR complex only by its binding to the radioactive PYR sequence γ-δ DNA in the gel shift assay; its physical association with PYR sequence defined the presence of PYR complex, nothing more.

We used as many antibodies as we could reasonably think of in an attempt to identify a PYR complex protein, but we failed to to do so. There was no GATA-1 or 2, no EKLF; there was nothing we could think of. We were groping for one sure component of the complex that we could then use as a hook to find others, but we did not succeed.

What we did discover relatively early was that the complex had a molecular weight of between one million and two million daltons (a measure of its size), suggesting that there were five to ten different proteins in the complex, but we had no idea what they were.

David used the tried and true way, referred to colloquially as "grind and find," to determine what proteins were in PYR complex by purifying the complex through several different chemical steps.

Since there was relatively little PYR complex even in MELC, we had to start with an enormous amount of these cells. We first purified PYR complex by chromatography based on its large size; next, we used its ability to bind to certain resins; then, we further purified it on the basis of its unique electrical charge; and finally, we utilized its DNA-binding properties: the strong affinity of PYR complex for PYR sequence oligonucleotides which we attached to a column resin bed. In this last step, we found the conditions under which PYR complex and little else stuck to the resin. We then pushed PYR complex off the column with a high salt-containing solution.

We monitored the purification at every step, again, with only our gel shift assay, and hoped that the steps we used didn't destroy the complex. Finally, David had a significant purification: a close to 1000-fold increase in the amount of PYR complex binding activity in the small amount of protein present in the purified final product.

But was our purification good enough for us to identify any of the proteins in PYR complex? To determine this, we took the most purified material we had isolated, our precious protein, and put it on a protein gel. We saw 12 to 20 protein bands of varying sizes. We had no clue as to what they were.

It was now 1998, seven years after we had first described PYR complex, and I was desperate to find some component in the complex that we could use to move forward. After a significant amount of personal pleading, I finally convinced Dr. Paul Tempst of Memorial-Sloan Kettering, an expert at mass spectrometry analysis (the most sophisticated technique available for specifically identifying specific proteins—even in tiny amounts), to use his skill to try to identify specific PYR complex components in the highly complex mixture on the gel.

To our great surprise and delight, he and his co-workers were successful. They found that at least four of the proteins in PYR complex were so-called "SWI/SNF" proteins, proteins known to be active in chromatin remodeling, the process that modifies the configuration of DNA-protein interactions in chromatin. SWI/SNF proteins are known to specifically activate genes in the region of chromatin at which they bind: they open up the relatively closed configuration of chromatin and allow specific genes to be expressed, much like a rose opens as it blooms.

SWI-SNF proteins are associated with histone acetylation and gene activation. As part of PYR complex, the SWI/SNF proteins might theoretically bind to the γ-δ intergenic region in adult red blood cells, and activate human β globin gene expression.

Tempst also found that PYR complex contained histone deacetylases (HDACs), proteins known to do the opposite of SWI/SNF: HDACs repress the expression of genes in chromatin by deacetylating their histones. They close the rose. The HDACs are themselves part of another repressive nuclear protein complex called NURD. As part of PYR complex, HDACs might repress γ globin gene expression in adult cells when they are bound to PYR sequence.

These results were completely in sync with our proposed mechanism for PYR complex action in adult red cells: a biologic activity that appears in adult-type red cells that could reconfigure the human β globin locus - turning off γ globin via the HDACs, and turning on β globin expression via SWI/SNF, a true globin switching complex.

We were very happy with these results, although no one else in the field seemed to notice, presumably because we had not supplied more details of the process, if indeed it occurred at all, as we hypothesized, in adult human red blood cells, *in vivo*.

Another problem with our logic about PYR complex being critical for human globin switching was that neither SWI/SNF proteins nor HDACs were known to bind to any specific DNA sequences, certainly not polypyrimidines such as those in PYR sequence. To have a more convincing story, we needed a specific DNA-binding component in PYR complex that anchored our complex to our γ-δ intergenic globin PYR sequence, and changed its configuration during human globin switching.

We were rewarded in this search in 1997 by a clue from a paper by Dr. Katia Georgopoulos who had characterized a unique protein called Ikaros, a transcription factor she had identified in human and mouse T lymphocytes. She also reported that Ikaros was present in a protein complex. Ikaros turned out to be the central piece of the PYR complex puzzle.

In 2000, David and our lab showed in an elegant (my word) study that PYR complex was, indeed, a single chromatin remodeling complex containing as its DNA binding subunit, Ikaros, a transcription factor, linked together with SWI/SNF and HDAC proteins. To prove this, David used a technique called immunoprecipitation in which he used an antibody to one component of the complex, in our case, against either Ikaros or a SWI/SNF protein, to precipitate the entire complex. When an antibody to Ikaros was used, HDACs and SWI/SNFs were also precipitated. Similarly, when an antibody to BAF 57, a SWI/SNF protein, was used, Ikaros was also precipitated. We had deciphered the major structure of PYR complex, and published the results.

Ikaros was the only component of PYR complex limited in its expression to adult blood cells, and, thus, we hypothesized that PYR complex could only form in cells that expressed Ikaros. SWI/SNF and HDAC components are much more ubiquitous (widespread) in the nuclei of cells. Thus, Ikaros is the rate-limiting protein in the formation of PYR complex appearing specifically in adult red blood cells.

But even with most of the components of PYR complex identified, we still had shown no link experimentally between human globin switching and PYR complex in animals; we only had circumstantial evidence of an association between them. To address this issue, we began to use so-called "transgenic" mice, in 1995.

Transgenic mice are mice that contain foreign genes introduced into their genome by experimental manipulation. As alluded to earlier, mice have no fetal genes, in contrast to primates and humans. They do have embryonic globin genes that produce embryonic globins that have similar functions to those of human γ globin. The embryonic globin genes are expressed in the mouse yolk sac at day 11 of fetal life (day 11PC or post-coitally). Then the mice "switch" to making adult mouse β globin, analogous to adult human β globin, in mouse fetal liver (a very active blood tissue) from day 13 on during gestation, and, later, in mouse bone marrow.

Transgenic mice had been genetically engineered by others to contain long stretches of DNA from the human β globin locus (including the human γ, δ and β globin genes and their intergenic sequences), and to express the human fetal and adult globin genes at appropriate times. Human γ globin is expressed in these transgenic mice in yolk sac cells at day 11PC, and human β globin predominates in mouse fetal liver (the adult hemoglobin-expressing tissue) at day 13PC and thereafter. We obtained these mice from Drs. Tariq Enver and George Stamatoyannopoulos from the University of Washington, who had characterized them, and we reproduced their results.

The first new experiments David undertook with these mice were to make our own new transgenic mouse lines in which PYR sequence was deleted so that we could determine if PYR sequence was involved in human globin switching. To do this, we deleted a 511 base pair intergenic γ-δ piece of DNA containing PYR sequence and adjacent nucleotides from the Enver transgenic piece of human DNA (Fig. 2, Appendix). These new transgenic mice were made with the help of Dr. Frank Costantini, a pioneer in the transgenic field and a colleague of mine for many years at Columbia University. It took David *et al.* about two years to make these PYR sequence-deleted transgenic mice.

We were interested in determining in these experiments whether these PYR sequence-deleted transgenic mice could switch, and, if so, whether removing PYR sequence modified the time of switching. David analyzed human globin gene switching in these mice by measuring how much human γ and β globin protein was made as they switched from producing human γ globin in the mouse yolk sac at day 11PC, to human β globin in the mouse fetal liver at day 13PC and later.

This was our defining experiment. We had worked for eight years on this project and we would either stand or fall on the basis of our results. We would have abandoned the project if these experiments did not show a role for PYR sequence in human γ to β globin switching in these mice.

I probably would not be telling this story if we had failed to obtain meaningful results. David found, to our mutual delight, that in the mice with PYR sequence deleted, there was a significant delay in human γ to human β globin switching, as compared to mice without the deletion. The results were clear and were published. The project had been successful in establishing a biologic role for PYR sequence in human globin switching in intact animals although it is possible that other sequences included in our deletion might also play a role.

In 1997, we had lost our NIH grant for research on human globin switching, presumably because of our slow progress, but in 1999, with the publication of David's results, we got back the grant, and we had the money and impetus we needed to continue further evaluating the role of PYR complex and PYR sequence.

Could we establish a biologic role for PYR complex as we had for PYR sequence in human globin switching in animals? That would be another important landmark validating our hypotheses. We had the opportunity to do so because the Georgopoulos lab had produced mice that made no Ikaros protein at all (Ikaros null mice). These Ikaros-null mice were ideal mice for us to use to answer the question: How important is Ikaros in our PYR story and in human globin switching? Dr. Georgopoulos graciously made the Ikaros null mice available to us.

The first thing we did with these Ikaros null mice was to see if they made any PYR complex at all. If we were right and Ikaros was the rate-limiting component of PYR complex, we should see no PYR complex in the blood cells of Ikaros null mice.

Rocio Lopez, a graduate student, did the first gel shifts with nuclear extracts from normal mice and Ikaros null mice, and showed that there was no PYR complex activity at all in the adult blood cells of the fetal liver in Ikaros null mice. This was a critical result validating our previous experiments and showing that Ikaros is required for PYR complex formation in animals.

Rocio also showed at that time that in mice heterozygous for the Ikaros mutation (one normal Ikaros gene and one null Ikaros gene), there was only about half as much PYR complex as in normal mice. We even had a dose-related effect of Ikaros on PYR complex production. We were thrilled. We could now use these mice to see what effect the lack of PYR complex had on human globin switching.

To do this, we had to create new lines of mice, this time by breeding the Ikaros null mice with the Enver mice (with the human γ, δ and β globin transgene) we had used in our earlier experiments. After establishing these new mouse lines, we then again analyzed human γ to β globin switching in mouse yolk sac, fetal liver, and in the adult mice.

In these Ikaros null mice, the human PYR sequence was preserved intact in the human β locus transgene, so we were only looking at the effects of not having Ikaros protein in these experiments. The results were conclusive. The Ikaros null mice with the human globin genes had delayed human γ to β globin switching; they had persistently more human γ globin in fetal liver at day 14 of gestation and later, than did normal mice.

We had confirmed the results of our previous transgenic mice experiments in which we had deleted human PYR sequence, and had proven that the presence of Ikaros and PYR complex was necessary for normal human globin switching. We had solidified the experimental basis for our view that both PYR complex and PYR sequence were necessary for optimal human γ to β globin gene switching.

We hypothesized that PYR complex and PYR sequence worked in the scheme of human fetal to adult globin switching as follows (Fig. 5, Appendix). In fetal cells without PYR complex, the human βLCR interacts with the γ globin gene to amplify γ globin expression, with other protein transcription factors and complexes facilitating these βLCR-γ globin gene interactions. When PYR complex appears in adult erythroid cells around birth, PYR complex, acting with other complexes and transcription factors, interferes with the binding of the βLCR to the γ gene, decreasing γ globin gene expression. In other words, the interference provided by PYR complex changes the configuration of chromatin at the β globin locus. The result of this reconfiguration is that the βLCR, blocked from binding to the γ globin gene by PYR complex, moves by default to the human β globin gene, thereby activating β globin expression (Fig. 5, Appendix). In addition, we hypothesized that activating components (SWI/SNF proteins) in PYR complex might increase human β globin gene expression as well.

Rocio published her results in 2002, and we were excited to do new experiments to study the details of this process further. New techniques, such as chromatin immunoprecipitation (ChIP) were available to study the details of how Ikaros and PYR have their effects. ChIP measures the precise interactions between DNA sequences and specific proteins in intact cells. Using this method, we might have found out precisely how and where PYR complex is bound to the human β globin locus DNA.

ChIP experiments had already been published by other groups showing that NF-E2 and GATA-1 were bound to both the βLCR and the β globin genes and were partially responsible for looping. Also, Grosveld had used ChIP technology to show that there was indeed a physical association by looping of the βLCR during switching, first to the γ globin gene in fetal life, and then, with switching, to the β globin gene in adult blood cells.

We had also started to use another technique, gene array analysis, to find out what genes and proteins were changed in their expression during globin switching in normal and Ikaros null mice. We were on the road to expanding our use of these new technologies, when suddenly, for reasons I could not understand, our NIH grant on this subject was not renewed, despite our significant (my word) progress in the field. Although I am still doing research at Columbia University and I still have an NIH research grant, it is to do experiments on human globin gene therapy, not globin switching.

More recently, in support of our hypotheses regarding PYR, Peter Fraser from Cambridge, England, a prominent researcher in the globin field, published experimental results in 2005 showing that in patients with relatively small deletions of human globin sequences that include PYR sequence upstream of the human δ globin gene, there is a change in the chromatin configuration in the γ-δ intergenic region, the region of the deletion; and that this change is accompanied by a marked increase in the activation of the human γ globin genes and in the production of fetal hemoglobin. These studies in humans again point to the importance of the PYR sequence region in human globin switching.

The PYR story continues today. New mouse transgenic models in which not only Ikaros expression but that of other Ikaros-like proteins are also inhibited have been described, and they show even more severe effects on blood production than we documented in Ikaros null mice. These so-called Plastic (Plst) mice die *in utero* from anemia at day 14PC. Data on these mice, recently published in the British Journal of Hematology by Drs. Janelle Keys and Andrew Perkins and their associates, show that "Ikaros drives human hemoglobin switching." In these experiments, they use ChIP assays to show that the lack of Ikaros changes the configuration of chromatin at the human β globin locus that favors the prolonged binding of the γ globin gene to the βLCR, and manifests itself in a significant delay in human γ to β globin switching.

I do not know whether research on PYR and Ikaros will ever contribute to a cure for Cooley's anemia. It might perhaps do so some day by using an anti-Ikaros approach (Ikaros inhibitors) to prevent PYR complex from forming in the bone marrow, thereby allowing continued γ globin production. Butyrate and other HDAC inhibitors, already available, increase fetal hemoglobin and may well act by affecting PYR complex activity in adult cells. As indicated earlier, combinations of drugs that increase fetal hemoglobin are also a potential treatment in the future.

In much of my research in hemoglobin synthesis and Cooley's anemia, I have usually been engaged in races to be the first of several investigators working in the same area to get a new result, usually soon confirmed by others. But because of our unique situation with PYR complex and PYR sequence in which our lab, until recently, has been the lone engine driving the research from the start, this research has been that much more scientifically challenging and rewarding to me personally. I was fortunate enough to be able to summarize all of our work on PYR complex and PYR sequence in a recent review article (2006) on the regulation of human fetal hemoglobin in *Blood*.

I have been excited to do the science and see progress in the understanding of human globin switching over almost five decades. I hope in the future there will be new destinations of success and meaning that can be reached in this important area that may lead to new ways to fully reactivate fetal hemoglobin production in the red cells of patients with Cooley's anemia, and thus, to a cure.

*   *   *

In summary, if we could fully reactivate fetal hemoglobin production in adult red blood cells, we could cure Cooley's anemia. We have described a complex of proteins (PYR complex) that has a role in normal human globin switching and in regulating fetal hemoglobin in adult red blood cells. PYR complex is an hematopoietic cell-specific protein complex that only appears in adult red cells, and is absent in fetal cells. It is composed of the transcription factor, Ikaros, and other more ubiquitous proteins known to change the configuration of human DNA in chromatin: some by opening up the chromatin and allowing new gene expression; others by closing the chromatin and shutting down genes.

PYR complex binds to DNA sequences (PYR sequence) between the fetal and adult genes at the human β locus, and may facilitate switching from fetal to adult globin production in adult red blood cells by repressing γ globin gene expression, and perhaps by activating β globin gene expression as well. PYR complex may function in adult red blood cells by physically blocking the interaction of the human β globin LCR with the γ globin gene, thus repressing γ globin production, and it may also help move the βLCR from the human γ gene region to the β globin gene region to activate β globin gene expression. In addition, the components of PYR complex may specifically activate human β globin gene expression directly.

The insights gained into the role of Ikaros and PYR complex in human globin switching may eventually be therapeutically useful by making it possible to increase human fetal hemoglobin to very high levels in adult cells. This approach to treating Cooley's anemia could result in a cure.

Part Three

The Best Medicine: Current Care and Future Goals

This section discusses the current treatment for thalassemia, novel approaches such as gene therapy and stem cell transplantation, and advances in antenatal diagnosis. It also illustrates the lives of two families with young thalassemia patients today.

Chapter 13

The Standard of Care

Blood and Desferal

The standard of care for patients with Cooley's anemia is better today than it has ever been. Blood transfusions to a hemoglobin level of nine to ten grams percent and iron chelation with Desferal sub-cutaneously via a pump constitute the current routine treatment for these patients. As Dr. Patricia Giardina, head of the thalassemia program at New York-Presbyterian Hospital at its downtown site, says, "Give me a patient who takes 60 milligrams per kilogram (mg/kg) of Desferal a night at least five nights a week, and I can assure you of negative iron balance. Essentially, I can prevent iron overload.

"This program is a miracle. In the US, I can give my patients as much Desferal as I consider appropriate and I can treat them optimally. I have all the luxuries of care and support for the patients, although there are still significant financial issues. My situation as a treating physician is different from that of those treating thalassemia elsewhere in the world. Especially those in third world countries. Even in Europe, the price of Desferal is dear, and the doses used are often lower and perhaps not as effective."

The optimal treatment program for Cooley's anemia also requires a safe and available blood supply. Over the past several decades, advances in ensuring blood safety have been made that have vastly reduced the most severe risks of blood transfusions, such as hepatitis and AIDS. The value and importance of a safe and available blood supply around the world for patients with Cooley's anemia cannot be overemphasized. Without adequate and safe blood, these patients will die, usually before the age of five.

Desferal is a drug developed by the pharmaceutical company, Ciba-Geigy (now Novartis). It was initially shown to remove iron from the human body when given by almost any route, except orally. Oral Desferal is not absorbed well and does not work, while intramuscular Desferal is somewhat effective, but painful. Intravenous Desferal is very effective, and is used to remove large amounts of iron in a short time, especially in emergency situations such as the development of cardiac arrhythmias by patients, but it is impractical for routine daily use.

Subcutaneous Desferal therapy is, in my opinion, the single most important advance in thalassemia treatment ever; it was introduced in the 1970s. After investigators in the US and Europe had shown that subcutaneous Desferal could lead to significant removal of iron from tissues, Drs. Richard Propper and David Nathan popularized this treatment, using a syringe attached to a small pump to deliver the drug slowly and continuously over 8–12 hours at least five times a week.

This prolonged exposure of the patient to Desferal is necessary because iron enters the blood stream relatively slowly from tissue stores and because Desferal has a relatively short half-life. The patient must stick a needle under the skin of the arms, legs, buttocks, or abdomen. The pump is attached to a syringe that must be held immobile; and sterility must be maintained. This difficult procedure must be performed almost daily, and is often painful as well. The original pump used by Propper *et al.* was designed to make the treatment as patient-friendly as possible.

Linda D. and Amy P., whom I have discussed, are examples of patients with Cooley's anemia who have had long-term success with frequent blood transfusions and subcutaneous Desferal, although not without the daily struggles accompanying this intensive therapy. Many other patients have succeeded as well. However, there are patients who have tried and failed to comply with this program, and have suffered the consequences. They have lost the struggle with the disease because maintaining the optimal Desferal treatment almost every day has not been possible for them. Compliance issues with subcutaneous Desferal constitute a momentous and devastating problem for these Cooley's anemia patients and their families that is difficult to resolve. Clearly, alternative therapies are needed, and Desferal is not a cure.

Oral iron chelators

Until recently, subcutaneous Desferal was the only approach to effective successful long-term iron chelation for thalassemia. The dream of all who have worked in the field has long been an oral iron chelator, as effective and relatively non-toxic as Desferal, that would free patients from the pain, trouble and problems involving the use of needles and pumps. Today, two new oral iron chelators, L1 and Exjade, are available for this purpose. L1 is approved for use in many countries around the world, but not in the US. Exjade has recently received FDA approval in the US.

I was introduced to a scientist named Dr. George Kontoghiorghes at a CAF Symposium that I chaired in 1990. He was giving a talk about the value of a new oral iron-chelating drug that he had developed in England in the late 1980s that was being tested in patients there by Dr. Victor Hoffbrand. The drug was called L1 (1,2-dimethyl-3-hydroxypyrid-4-one, or Deferiprone), and he was proselytizing from the podium for its use.

He generated enthusiasm with his presentation, and a lively discussion of his results followed; but I finally had to tell him that the discussion time for his talk was over and that we had to move on. Well, he did move on, scientifically and politically. He was truly the prime mover promoting the use of L1, especially in England in the late 1980s.

Dr. Nancy Olivieri, a Canadian physician, joined in using L1 in 1990. She showed, in early studies at the Hospital For Sick Children in Toronto, that L1 was effective in removing iron, and she helped persuade Apotex, a drug company in Canada, to make and market the drug. The original chemical formulation of L1 had never been patented, but the British government had licensing rights for human use of the drug. Ciba-Geigy (now named Novartis) had initially tested the drug in animals, but abandoned its development in the 1980s, citing its toxicity.

Apotex eventually acquired the rights to use L1, and began making the drug. The company, the largest pharmaceutical company in Canada, had previously made and sold generic versions of approved drugs. Apotex was relatively inexperienced in developing and testing new drugs; in doing extensive testing in animals for toxicity; and in doing the kind of orderly clinical testing required to assure US FDA approval.

In the 1990s, there was a falling out between Olivieri and Apotex that impeded L1's clinical availability. Questions about its relative safety and efficacy arose that led to delays in its use in the US and Canada. Olivieri reported severe liver toxicity in patients receiving the drug in her clinical trials in Canada which, until then, had been funded by Apotex. Olivieri and Apotex parted company and she vigorously campaigned against L1's continued use in Canada and worldwide, citing patient safety.

However, concurrent and subsequent larger clinical trials in Europe did not confirm the liver toxicity that Olivieri had reported. Moreover, L1 has since been shown to be effective in removing iron in thalassemia patients, and has been life-saving in patients who cannot tolerate Desferal or cannot afford or obtain it. It is now being used widely throughout the world, primarily in Europe and Asia, but is still not approved for general use in the US or Canada.

A detailed account of the development of L1, and the controversies surrounding its use, is presented in the book "Drug Trial: Nancy Olivieri and the Science Scandal at the Hospital for Sick Children" by Dr. Miriam Shuchman, a psychiatrist and Canadian journalist. I recommend it to anyone interested in the subject.

L1 is approved for use in the European Union. It is also widely used in India and other countries in which Desferal is not a viable alternative for many patients because of the cost of the drug, and the lack of availability of qualified medical personnel and facilities to monitor its relatively complex use.

In the late 1980s, having abandoned L1, but with new indications that it might succeed Desferal as the iron chelator of choice, Novartis (previously Ciba-Geigy) developed a new oral iron chelator of its own. The company was in danger of losing its financially lucrative market for Desferal, estimated at hundreds of millions of dollars a year, to L1, the cheaper and much easier to use oral alternative. A pharmaceutical giant, with extensive experience in new drug development, Novartis relatively quickly found a new effective oral iron chelator, Exjade (ICL670A, deferasirox), that recently obtained US FDA approval.

Exjade is currently the only oral chelator available in the US, and is being widely used as an alternative to Desferal throughout the world. It is very expensive compared to L1. L1 is also cheaper than Desferal, and L1 plus a lower dose of Desferal is being extensively studied in Europe as a cheaper alternative to higher doses of Desferal alone; it has not yet been proven that L1 alone is as effective as Desferal.

There is no question that L1 and Exjade have been life-savers for many, many patients throughout the world for whom subcutaneous Desferal and the pump are impractical. However, long-term experience with Exjade and L1 and their risk versus benefit profiles are necessary to determine whether they can save thalassemia lives over decades as safely and effectively as subcutaneous Desferal has been shown to do.

All of these drugs have side effects. Desferal is known to cause cataracts, rashes, and other assorted side effects, but it has remarkable safety. Very few deaths have been attributed to it over its decades of use. Exjade causes rashes, and recently has been shown to be associated with renal toxicity. L1 can cause arthritis; it can also cause low white blood cell counts, and deaths from infection have been reported. However, with proper laboratory monitoring and appropriate medical intervention, the potentially dangerous toxicities of these relatively new oral drugs can likely be avoided.

Bone marrow transplantation

Some patients with Cooley's anemia have another choice in fighting their disease, allogeneic bone marrow transplantation (ABMT), which is currently the only treatment that can result in a cure. This procedure involves the transplantation of blood stem cells from a person with normal red blood cell production into a patient with Cooley's anemia. Dr. Giardina says, "If I have a new thalassemia patient who meets the criteria for ABMT, I offer them that option. These are usually younger patients with suitable donors. This is an established therapy that works."

Dr. Alan Cohen, Physician in Chief at Children's Hospital of Philadelphia (CHOP) and an expert in thalassemia, agrees. "As long as the transplant center is a good one with lots of experience, transplantation is a reasonable option," he says, adding, "For some reason, marrow transplantation is a more widely accepted treatment in Europe, especially Italy, than in the US. I'm not sure why."

Dr. Guido Lucarelli, an Italian transplanter, was a pioneer in this field in the 1980s, and that may be one reason. Lucarelli did his first transplant in Pesaro, Italy, in the early 1980s, soon after Dr. Donnall Thomas did the first transplant ever in a thalassemia patient, in Seattle. In the US, Dr. Mark Walters, a transplanter at Oakland Children's Hospital, is one of the most experienced in using this procedure to treat thalassemia, and he and Dr. Elliot Vichinsky, current Medical Director of CAF, have been prime movers in this field in the US recently.

Our bone marrow produces all of our red blood cells from remarkably few primitive nucleated cells, called blood stem cells. These stem cells are the only blood-forming cells that can both divide to make more of themselves, and, through differentiation, also develop into all of the normal blood elements. Thus, these few blood stem cells lead to all of the nucleated red blood cells in the bone marrow and, in turn, all of our hemoglobin-containing circulating red blood cells. Blood stem cells produce white blood cells and platelets as well.

A single blood stem cell can repopulate the entire blood forming system of a mouse whose bone marrow activity has been totally eradicated by drugs or irradiation. In humans, only a few hundred stem cells are required in this situation. The power of blood stem cells as biologic factories capable of rescuing the blood system of animals, including humans, is truly awesome.

Successful transplantation of the blood stem cells themselves is required for the long-term cure of thalassemia. By contrast, any later red blood cell precursors transplanted that are differentiated beyond the stem cell stage cannot substitute for true stem cells. These later cells simply differentiate into more mature red blood cells that are destined to die after weeks to months, and are not useful for cures.

Successful bone marrow transplantation solves the only medical and biological problem in patients with Cooley's anemia: the presence of genetically defective red blood cells which cannot make enough normal adult hemoglobin, HbA ($\alpha_2\beta_2$). With ABMT, the patient's bone marrow is replaced by normal blood stem cells which can produce enough normal mature red blood cells with normal amounts of human β globin and HbA. If successful, ABMT cures the anemia and eliminates the need for transfusions. Thus, there are no toxic consequences from excess iron, and no need for chelation. The patients are cured by a single procedure.

Until a decade ago, the only source of blood stem cells for ABMT was the bone marrow. At that time, it was found that by administering certain drugs which stimulate their migration from bone marrow into the blood stream, blood stem cells can also be obtained for transplantation from circulating blood. It has been demonstrated that these so-called peripheral blood stem cells (PBSC) are as effective as marrow-derived stem cells for successful transplantation. ABMT using PBSC and bone marrow-derived stem cells is often referred to as blood "stem cell therapy" or SCT. These blood stem cells should not be confused with multipotent embryonic stem (ES) cells which are derived from early embryos.

PBSC harvest can be done relatively safely, and simplifies stem cell donation since it does not require putting the donor to sleep for the more painful and tedious harvest of stem cells from the bone marrow. Multiple marrow sticks, primarily in the pelvis, are necessary for routine bone marrow harvesting for ABMT. Stem cells from umbilical cord blood have also been used for ABMT.

Blood stem cells obtained from any source can be given to patient recipients by vein (intravenously), and, as if by magic, they migrate directly to the recipients' marrow where they grow and differentiate. This process, called "homing," occurs relatively efficiently because the blood stem cells have specialized protein molecules on their surface that interact with complementary molecules (receptors) on other marrow cells that attract the stem cells to the marrow and hold them there.

So today, ABMT or SCT can be done relatively easily by harvesting blood stem cells from a vein of the donor, and infusing them intravenously into the recipient.

There are, however, several problems in doing ABMT. One is that the bone marrow of the recipient patient or host must be cleared of its own defective stem cells to allow the donor's marrow to "take" or seed. Without removing all of the recipient's remaining marrow stem cells, the donor stem cells cannot park in the marrow. This situation exists because the donor cells have no advantage over the thalassemia or other cells already parked in the host's marrow. Like spaces in a parking lot, the number of niches or places in our bone marrow for stem cells are limited. In ABMT, we need to move out all of the cars already parked there, before we can park any new cars there.

Space for the curative donor cells is made by killing or ablating the recipients' marrow stem cells by drugs and/or irradiation. This process, also called "conditioning," is intrusive and dangerous, but is required for the success of SCT. Because of the conditioning process, the recipients have very low levels of all blood cells for one to two weeks, or more, before the establishment of the new normal marrow by SCT. However, today, most patients recover relatively quickly, uneventfully and completely, supported by red blood cell and platelet transfusions, and antibiotics to prevent or treat infections, until the donor stem cells function adequately.

In modern day bone marrow transplant centers, extra donor marrow or PBSC is obtained at the time of original harvesting, to be used if the initial dose of donor stem cells given is inadequate. The risk of death from failure of the donor marrow to engraft is relatively low for this dramatic procedure (wiping out and reconstituting an entire human organ system). Today, while the risk of death from ABMT varies from center to center, it is quite low, easily less than 5%.

The most serious problem with SCT is immunologic incompatibility. We each have antigens, called human leukocyte antigens (HLA), on the surface of our blood cells, including our stem cells, that predictably lead to immune reactions if they are not the same in donor and host. HLA incompatibility leads to both the production of antibodies to the newly introduced donor cells, and the generation of new immune-type cells, T (thymus-derived) lymphocytes. HLA antigens are inherited from each of our parents. We call the compatibility of the recipient's HLA antigens with the donor's HLA antigens histocompatibility.

Histocompatibility is required for successful SCT. Without complete histocompatibility, a unique and potentially devastating process called graft-versus-host disease, or GVH, can be initiated. GVH can lead to chronic skin, blood vessel, and liver problems, and can, occasionally, be fatal.

The HLA antigens in each of us are varied, but limited in their number. We inherit these antigens, as groups of molecules, through genes that are linked to each other in our chromosomes. Identical twins have the same HLA antigens. Statistically, there is a one in four chance in any family that any one child will have the same HLA antigens as any other sibling.

When histocompatible donors, usually siblings, are used to transplant Cooley's anemia recipients, over 90% of the patients are cured by the procedure, and have little or no GVH. That is remarkable! Younger patients and those with fewer transfusions and no liver disease are preferred for ABMT. Older patients and those with a greater number of transfusions, or with liver disease, do worse.

Less well matched donors, donors that are haploidentical (half-identical) or those with limited or even minor mismatches, are more problematic. Additionally, less than complete marrow ablation in ABMT is a highly desirable goal because it would be less toxic. However, it has thus far been largely unsuccessful as conditioning for Cooley's anemia. Clearing the parking lot only partially of cars is not a current option.

Clearly, ABMT in one treatment can lead to a cure of Cooley's anemia in most patients who have histocompatible donors. This is an extraordinary outcome: the cure of Cooley's anemia for thousands of patients worldwide.

A major limitation of ABMT or SCT for thalassemia, as with other diseases, is that only a relatively small percent of patients, less than 30%, have suitable completely histocompatible donors. However, the availability today of *in vitro* fertilization (IVF) and pre-implantation screening of embryos, in this case to find potential siblings who are HLA identical to thalassemia patients, may dramatically increase the number of patients whom ABMT can cure. In this process, embryos generated *in vitro* in a tissue culture dish are screened for those with complete histocompatibility with the patient. Only HLA-identical embryos are implanted into mothers or surrogates so that they will give birth to siblings who are optimal potential donors.

The great hope for patients with Cooley's anemia who do not have appropriate histocompatible donors is human β globin gene therapy. The specific advantage of gene therapy is that, if successful, this procedure would be available to all patients with Cooley's anemia since the corrected blood stem cells are those of the patient alone, and, therefore, there should be no immunological problems with this approach. The curative potential for gene therapy in thalassemia will be discussed in a later chapter.

Drugs increasing fetal hemoglobin

Another approach to the cure of Cooley's anemia, as I've discussed previously, is to reactivate the production of human fetal hemoglobin to very high levels. Thalassemia patients have normal fetal globin genes that are maximally active in producing fetal hemoglobin (HbF, $\alpha_2\gamma_2$) *in utero*. Can we fully reactivate these fetal genes with drugs and cure the disease? Not yet.

Drugs that increase fetal hemoglobin have been studied intensively over the past two decades with only modest results. Hydroxyurea and derivatives of butyric acid have been the agents primarily used. Both types of drugs have had some positive results in clinical trials, however, in most patients, only modest increases in fetal hemoglobin are attained. Other agents have also been tried, but, again, with only limited success.

Dramatic increases in HbF using derivatives of butyric acid have been reported in individual patients, but the overall track record of this treatment is quite mixed. There appears to be a wide variation of individual responses to these agents, perhaps due to genetic factors.

Dr. George Atweh, an expert in the field from Mount Sinai Hospital in New York, said in an article in 2005, "Although hydroxyurea was approved by the FDA for the treatment of sickle cell disease in 1996, no similar pharmacological agent(s) has been approved for the treatment of patients with thalassemic disorders. The small-scale studies of the induction of fetal hemoglobin in thalassemia have been generally disappointing."

Dr. Susan Perrine, the discoverer of butyrate and a pioneer in this area, is an advocate of combining several novel agents together for treatment, for example, butyrate with other compounds. Erythropoietin plus butyrate has been tried, but, again, with only limited success. New fatty acid derivatives that Dr. Perrine and others have discovered may also be useful.

The hydroxamic acid derivatives of short-chain fatty acids, butyryl and propionyl hydroxamate, subericbishydroxamic acid, and suberoylanilide hydroxamic acid, are potent inhibitors of histone deacetylase (HDAC), as is sodium butyrate. They have been shown to induce fetal hemoglobin in tissue culture systems, and in transgenic mouse models. Whether they will be useful in human clinical trials is unknown at present.

Dedicated Doctors

Blood transfusions became routinely available after World War II. The needs of the wounded on the battlefield had spurred the growth of technology for obtaining safe and available blood, including the use of ABO typing. Dr. Carl Smith became the head of Pediatric Hematology and Oncology at New York Hospital (NYH) in 1942, and started a thalassemia program there after the war. According to Dr. Patricia Giardina, the current chief of the Division of Pediatric Hematology and Oncology, Dr. Smith originated the concept of an out-patient transfusion clinic and created the Children's Blood Foundation for those needy patients who required blood transfusions and specialty care that otherwise would have been prohibitively expensive. "The mission of the Children's Blood Foundation was one I completely embraced and continue to even today. [Its goals] are to provide 'outstanding clinical care' and to 'foster research to advance knowledge of disease in order to improve treatments and generate curative strategies.' " Dr. Giardina is head of the Foundation today.

Dr. Smith founded the pediatric thalassemia program at NYH which has become the top clinical care facility of its kind in the New York metropolitan area. Many outstanding pediatric hematologists have directed the thalassemia program under Dr. Smith's tutelage, including Drs. Marion Erlandson, Gertrude Stern, Virginia Canale, Alice Jane Markenson, and Patricia Giardina who came to the Cornell-New York Hospital residency program in 1969 and met Dr. Smith at that time. She completed her training at NYH (now New York Presbyterian Hospital-Weill Cornell Medical Center). Currently, in addition to being director of the thalassemia program at NYH, Dr. Giardina is also the Principal Investigator of one of the Clinical Research Centers in the North American Thalassemia Clinical Research Network (TCRN) which is funded by the NIH and the Centers for Disease Control and Prevention (CDC) for the prevention of complications in thalassemia.

When Dr. Giardina became thalassemia clinic director, the iron chelator, Desferal, had just gone into clinical use. "We all worked hard to find the best regimens to give Desferal. How much should we give? When should we start it? Should we change the dose as the patients grow up?" Dr. Giardina has been involved continuously over many years in trials and experiments to try to answer these questions which are so important, since preventing iron deposition in organs is critical to the clinical outcomes of Cooley's patients.

"It was found that starting Desferal chelation in infancy stunted growth. Some Cooley's doctors felt that the short stature of the patients was a necessary trade-off to prevent iron toxicity to the heart and liver in these very young children. Our studies showed that starting patients on Desferal at age four to five could avoid the short stature, and still get adequate chelation without residual iron overload."

Another major problem in Cooley's anemia is normal development of puberty. Iron deposition in the pituitary-adrenal axis is thought to delay or prevent pubertal changes in many Cooley's patients. Dr. Giardina and her colleagues looked carefully at the effects of various schedules of Desferal on iron balance and sexual and bone development. She and others found that higher doses of Desferal were required for some Cooley's patients to have normal growth and development.

In the '60s and '70s, estrogens and androgens were thought to be dangerous owing to their potential long-term side effects, but now they are used routinely if needed, to induce puberty. Similarly, human growth hormone is now often used to promote normal skeletal development in Cooley's patients. Because of the clinical studies of Dr. Giardina and others, most Cooley's anemia patients today are indistinguishable physically from their normal siblings. Dr. Giardina points out that these hormonal therapies have not only had an enormous positive effect on the physical health of the patients, but have also been a boon to their emotional strength and self-esteem.

Dr. Giardina also thinks that, in general, females have less severe Cooley's anemia than males. She cites anecdotal data from families in which brothers and sisters, with the same thalassemia genes leading to Cooley's anemia, have had significantly different outcomes. If these gender differences exist, the reasons remain unknown.

There are other parameters or genetic modifiers that can theoretically lead to variations in the severity of the disease among different patients with Cooley's anemia. One obvious one is that different individual thalassemia gene mutations *per se* are responsible for these variations. For example, patients with two mild β plus thalassemia genes can have less anemia, although most patients with combinations of either β zero or β plus genes have Cooley's anemia requiring life-long blood transfusions.

There is a clinical entity called "β thalassemia intermedia," defined as a condition in which a patient with β thalassemia has a mild to moderate anemia and does not need to be regularly transfused. This condition may exist because the patient has two "mild" β plus thalassemia genes. Less often, it is caused by the presence of a single "severe" thalassemia gene. The inheritance of one β thalassemia gene sometimes occurs with the concomitant inheritance of more than the usual number of α globin genes that lead to an increased excess of α globin. In this case, the result can be some anemia with stable hemoglobin levels of seven to eight. This situation is in marked contrast to that of little or no anemia in most people with β

thalassemia trait. Other than too many α globin genes, confounding factors not yet known may also cause thalassemia intermedia.

The major issue posed by patients with thalassemia intermedia is whether and when to transfuse them. If their hemoglobin levels remain relatively stable, above seven, is transfusion necessary? This issue is still largely unresolved.

There is no question that the overwhelming number of Cooley's anemia patients with two β thalassemia genes require transfusions. Their hemoglobin levels will continue to fall and endanger their lives without transfusion. The important unanswered question in these cases is: What causes the different outcomes in different patients with Cooley's anemia? Or, why do some of these patients live into their 40s and 50s like Linda D. and Amy P., and others do not?

Compliance with iron chelation therapy and the maintenance of negative iron balance is certainly one important prognostic determinant. As stated previously, Dr. Giardina believes that treatment with prolonged infusions of 60 mg/kg of Desferal almost every day reduces iron overload 100% of the time.

But compliance with treatment is not the only factor in determining prevention of iron overload and survival. Some compliant patients succumb to iron toxicity, and some non-compliant patients do relatively well long-term. One explanation for these different outcomes is that different patients with Cooley's anemia handle iron loads differently. Some are more likely than others to accumulate iron at the same level of chelation and transfusions.

Although the role of many different genes in normal and abnormal iron metabolism has been elucidated in recent years, relatively little is known about how these genes function in patients experiencing large iron loads due to transfusions, as is the case in thalassemia. Even in the inherited disease, hemochromatosis, in which iron overload occurs, the mechanisms have not been fully elucidated.

Increased absorption of dietary iron by the gastrointestinal tract occurs in thalassemia even without transfusions. This is a paradoxical effect somehow related to the increased iron turnover in these patients. How is this absorbed iron and that from transfusions handled in different patients? Is there organ-specific deposition of iron due to different genes in different patients that determines, for example, the extent of cardiac disease in thalassemia? What iron-modifying mechanisms are we missing in our research in this area in thalassemia?

Ineffective erythropoiesis, the process of increased production and destruction of abnormal thalassemia cells in the bone marrow of patients with thalassemia, also needs to be better understood. Dr. Stephano Rivella, who is working with Dr. Giardina, is studying this process, and has suggested that genes and proteins involved in iron metabolism, and their effects on osteoclasts and osteoblasts in bone remodeling, may be important modifiers of iron metabolism in different thalassemia patients.

For a long time, I thought that the bone abnormalities due to ineffective erythropoiesis were primarily due to too many cells being made in the thalassemia marrow, mechanically eroding and remodeling bones, and that was it. But the effects of ineffective erythropoiesis in Cooley's anemia are clearly more complex and varied than that: cell death or apoptosis (the process by which cells die); bone resorption and abnormal bone growth; inflammatory cytokines and other protein products, all play roles in these bone abnormalities and may play additional roles in the way different thalassemia patients handle iron.

Dr. Giardina has about 120 patients with Cooley's anemia under her care. We talk in her small office on the sixth floor of the Payson Pavilion at New York Hospital. There are several open cubicles with record keepers, nurse-practitioners, and nurses on the busy floor. There is a large bright waiting room, and an infusion center in an even larger area, with many comfortable chairs arranged in a semi-circle for patients and for family members as well. It seems like a homey atmosphere.

Dr. Richard O'Reilly, an expert in marrow transplantation, runs a major bone marrow transplant center across the street at Memorial

Sloan-Kettering Cancer Center. Dr. Giardina says, "I have about 19 or 20 patients who have had bone marrow transplantation with Richard over the past 25 years. Since I have about three new thalassemia patients on average a year, that means that about 25 percent of my new patients over this time have been cured by allogeneic BMT. I offer all my patients, especially those younger than age 12, allogeneic BMT, and they usually accept it, if they are eligible with suitable donors. If not, they go on the current transfusion and iron chelation programs.

"And now there's Exjade, and soon, maybe, L1 will be approved in the US. We are also evaluating other oral iron chelators and combinations of these chelators. The Europeans don't like to use as much Desferal as we do because it is expensive, so they favor combinations of L1 and lower doses of Desferal to maintain negative iron balance. I have used higher doses of Desferal because I know they work and because L1 is not FDA approved in the US."

Dr. Giardina emphasizes, "This is all quite different from 30 years ago. There are so many more regimens and options than before, and some day perhaps we will have gene therapy available to cure all of our patients.

"Also, worldwide, there are really two kinds of treatment for patients with Cooley's anemia," she adds, "that in the US and other developed societies, and that in the rest of the world. I can get almost everything I need and want for my patients. It's wonderful. But that's not true elsewhere."

Dr. Patricia Giardina is an outstanding caretaker and clinical researcher, dedicated to her patients.

Chapter 15

Human Beta Globin Gene Therapy

Making new hemoglobin: uphill all the way

It was a dream come true for me. I had wished hard for that day in 2006. The first patient treated with globin gene therapy for Cooley's anemia in our Paris gene therapy trial was expressing some of the new normal hemoglobin we had given her. "That's great," I said when I heard the news. A chill ran up my spine. The low percent of the new hemoglobin she was making wouldn't cure her, but it was a positive result. We had reached a milestone, still relatively small, but absolutely vital to ultimate success.

This was the first time any clinical investigators had ever seen new normal hemoglobin expression from an added hemoglobin gene in a human being, using human β globin gene therapy. How long would it last? Would it change or even go away? We didn't know the answers to these questions, but the news that she was really making new hemoglobin from the new human β globin gene we had given her was a positive sign. In science, and especially in clinical trials, you don't usually hit a home run in your first few times at bat. We had gotten to first base.

Human gene therapists have been up at the plate many times before, with very few hits and lots of strikeouts, and this looked like a hit. I have been doing science and hematology for over 40 years, and gene therapy research and clinical trials for the past 20 years, and this was a promising result.

I knew that this would be good news for the Cooley's anemia community, especially for the parents of patients with the disease. Cooley's anemia, with few exceptions, eventually kills its victims, either early in life or, with recent advances, in their 30s or 40s. I had recently talked to Nunzio Cazzetta, the parent of two children who had died with the disease, and he was particularly excited. "I know you guys have been trying a long time. I hope the cure is finally here," he told me. On the one hand, I felt guilty that, once again, I might be raising false hopes of a cure in the near term. On the other hand, I thought that the time to try had come. How long can you wait before you try? Every day is a lost day for some of the patients with the disease.

Frank Somma, the President at that time of the Cooley's Anemia Foundation (CAF) had told me again and again in recent days, "We want a cure. We'll do anything we have to do to get one. Just tell us what you need." The CAF is the main advocacy group for Cooley's anemia patients and has been for the past 50 years, and I have been associated with it for most of that time.

I know how deep their commitment to finding a cure is, and that they would do anything to have it happen as soon as possible, especially with so many of their children now in their teens and 20s, and always facing the dangers of more severe iron overload, despite chelation, and sudden death due to heart disease in the years to come. Frank and the other Cooley's parents are some of the most dedicated people I have ever known, and I wished for a cure for their sake, and their children's sake most of all.

Human gene therapy for thalassemia is theoretically simple. Take blood stem cells from the patient, add a normal hemoglobin gene to the cells outside the body, and return the gene-corrected cells to the patient. I had prayed hard for an uneventful recovery of the first patient, whose initials are TK, a 26 year old woman, from the gene therapy procedure in Paris, even before hearing the good news about her new hemoglobin expression.

We had been thinking about the possibility of doing gene therapy for thalassemia since 1972, when we discovered how to make β globin cDNA. Way back then, we thought we could use the cDNA itself as the source of the normal human β globin gene sequences that could cure the disease. In fact, in the 1970s, others had tried to do just that. It failed badly, and the American investigator involved in this trial was severely reprimanded by the NIH for not following the prevailing rules for ethical clinical investigation.

It became clear early on that, in addition to the coding sequences in the cDNA, other important regulatory elements, such as the intervening sequences within the gene, and regulatory sequences upstream and downstream of the human β globin gene, are all required for successful and high level human β globin gene expression. Dr. Frank Grosveld's discovery in the late 1980s of important regulatory sequences far from the β globin gene itself, called the β locus control region or βLCR, was seminal in moving the field of β globin gene therapy forward. The βLCR is key to providing a dramatic increase in the human β globin gene's output of human β globin, and this increase is necessary for the success of the gene therapy approach. (Selected relevant references to human β globin gene therapy are provided in the Appendix.)

Clinical trials of human gene therapy for other diseases have been fraught with unforeseen tragedies in recent years, and so, on November 16, 2006, I was gratified to hear that TK had gone home, after 60 long, hard days in the hospital. She had survived the bone marrow gene transplantation part of the procedure, the most dangerous part, and was relatively well, and that was the first important clinical result of this new experimental protocol. She had undergone a long physical ordeal to try to be cured of the disease that had plagued her all of her life, and that had led to many previous hospitalizations for disease complications.

TK had received many blood transfusions and iron chelating agents and despite these treatments, she still had iron overload and was in danger of sudden death from the cardiac toxicity of the excess iron in her transfusions. Gene therapy was a possible alternative therapy for her.

Gene therapy as a treatment approach in humans in recent years has been made more credible by the fact that it is possible to cure mice with the equivalent of human β thalassemia by adding a normal human β globin gene to their blood stem cells; the thalassemia mice produce large amounts of normal human β globin. But it is a long way from mice to humans.

Blood stem cell gene therapy

It has been especially difficult finding a safe and efficient way of transferring a normal human globin gene into the blood stem cells of patients, like TK, and having those cells make the new normal hemoglobin needed to survive. On the other hand, human marrow gene therapy has long been a feasible way of getting genes into blood cells, including blood stem cells, without affecting all of the other cells of the body. You take the cells you need from the blood or bone marrow, put in the corrective human β globin gene, and then give back the corrected cells by vein, after which the transplanted blood stem cells know how to find their way back to the bone marrow without help! The process is simple enough in theory, but applying it to human patients, like TK, had turned out to be terribly complicated.

Blood is made in the marrow cavity in our bones. TK had her marrow blood cells removed two months before; then normal human β globin genes had been inserted into the collected cells, and the cells, hopefully gene-corrected cells, had been returned to her by vein.

We had taken a great risk in this first human globin gene therapy clinical trial by completely destroying (ablating) TK's own bone marrow, both to rid her of all of her own defective blood stem cells, and to provide marrow space for her gene-corrected cells to flourish. Every clinical protocol is a compromise between potential benefit and perceived or known risk: the so-called benefit-risk ratio. In this protocol, we opted for the increased risk of full marrow ablation to increase our chances for a meaningful therapeutic benefit from the transplanted gene-therapy corrected cells.

We elected to do this because previous experience in human allogeneic bone marrow transplantation (ABMT) has shown that less than complete ablation of a recipient's bone marrow is usually insufficient to allow successful transplantation of donor cells. On the other hand, ABMT with complete marrow ablation, although known to be very successful, can also be more dangerous, because it takes a longer time for the patient to recover from its major toxic side effects: a decrease in white blood cells which increases the possibility of infections, and a low platelet count which increases the danger of bleeding.

We thought the potential advantage of complete ablation of the patient's marrow outweighed the increased risks and gave us a much better chance at successful transplantation of the gene-corrected cells. But we were scared. This was a dangerous journey. Now I had heard that TK had survived several weeks of very low white blood cell and platelet counts. She had, of course, been supported by antibiotics, white cell growth factors, and platelet transfusions to ameliorate these complications, and she had also continued to receive red blood cell transfusions to combat her anemia. But, thankfully, she had now recovered her marrow function.

The use of complete marrow ablation in our study was by no means universally looked upon by our fellow scientists and ethicists as an acceptable risk. But as one of the ethicists had noted at a conference on the subject, "There is nothing unethical about increasing the level of risk in adults if they have given informed consent, and if you are convinced there will be no benefit without that risk."

The final arbiter in approving these protocols in the US is the FDA, and similar groups exist in other countries. We had decided to do our first clinical trial in Paris since most of the investigators we wanted to work with were there. These investigators had experience in the field of human gene therapy, especially Dr. Marina Cavazzana-Calvo, who has done extensive human clinical gene therapy trials.

We went through the French equivalent of the FDA to obtain permission to do the study. The protocol's review by the French agencies was extremely detailed and thoughtful. It took an additional year, after we were ready to do the trial, to satisfy their concerns and we received approval in the fall of 2006.

TK's gene therapy treatment, however, had not been ideal since, despite our best efforts, she received only one-third as many gene-corrected cells as we had optimally planned to give her. However, we transplanted her with all of the gene-containing cells available to us.

Our magic medicine

While naked hemoglobin DNA in some form could be used as the "vector" (or carrier) to transfer and express the new normal β globin gene, it is an inefficient way to do gene therapy. How do we transfer our gene into the patient's blood stem cells? We do this by using pieces of viruses, so-called defective viruses that, unlike normal viruses, are unable to reproduce themselves after they have inserted their genetic material into our chromosomes. All our β globin virus vector can do is transfer our globin gene sequences into the blood stem cells, integrate our globin gene into chromosomal DNA and allow the expression of the potentially curative human globin gene.

Viruses are simply pieces of RNA or DNA wrapped in specialized viral proteins that allow the viruses to enter cells and reproduce more of themselves. They then use proteins in our cells to make more of their own proteins after they have infected us. Their goal is simply to produce more viral particles inside our cells.

Once inside our cells, RNA viruses use an enzyme, reverse transcriptase (RT), to make a DNA copy of their RNA, and integrate this genetic material into our chromosomal DNA. After this, two things can happen: either the viruses stay there harmlessly, or they produce more viral particles, kill our cells, release these viral particles and infect more cells as they march on. This whole process seems quite wasteful and useless from the standpoint of human evolution, but these viral nucleic acids and proteins predate us, and have been extensively involved in our evolution.

In fact, these viruses have some of the same goals as ours in gene therapy: to have new genes efficiently enter our host cells, integrate into our DNA and produce the encoded proteins. What we do in our gene therapy is to manipulate and use parts of viruses to help us encode and transfer our new human β globin gene into cells, and to express our new human β globin protein. We arrange it so that the viral pieces we use cannot, by themselves, make any new infectious viruses. They are defective viruses. They are simply our missiles for transferring genes and having them expressed in blood stem cells, nothing more.

Some viruses just stay in the cytoplasm of our cells and don't enter the nucleus. These viruses are not useful to us. For human blood stem cell gene therapy, we need to use viruses that can transfer our added globin genes from the cytoplasm into the nucleus of stem cells, and allow our genes to integrate into the patient's chromosomes so that when the gene-corrected stem cells divide, our new genes are transferred and maintained in all daughter stem cells. The ideal viruses to accomplish this are called retroviruses.

Twenty years ago, I thought it would be a cinch to use defective retroviruses to do human gene therapy. I envisioned the treatment as a pill containing a virus with the globin gene inside that uncoats in the stomach, enters the blood stream, exits at the right tissue location, in our case, the bone marrow, and does its work. Hepatitis virus certainly knows how to target specific cells, only liver cells; AIDS virus targets only T lymphocytes; and there are many other viruses that only enter specific types of cells. But it has not been possible to harness the tissue-specific functions of these potent viruses for our globin gene therapy since there are no viruses available currently that specifically target blood stem cells.

Instead, we have been forced to use viruses that are not tissue-restricted and have the capacity to enter most types of cells. But since we can obtain blood stem cells from either bone marrow or circulating blood, we can remove these cells, and add our globin gene-containing retrovirus outside of the body (*ex vivo*). Thus, we have a great advantage in this respect over other gene therapy approaches where the target organ cannot be removed. We limit exposure of the virus to the gene-targeted blood cells we use. We are not in danger of affecting non-blood cells by our treatment, as we would be if we were trying to get corrective genes into, for example, the brain or heart or kidney. Additionally, we might not be able to achieve the high levels of gene transfer we desire without toxicity if we had to give our gene therapy virus *in vivo*.

We have found through trial and error over the past two decades that we can only utilize certain types of retroviruses for human globin gene therapy. We and others have previously used so-called gammaretroviruses (γ retroviruses) in human clinical trials to transfer and express human genes, including the human β globin gene in blood stem cells. In the 1990s, Dr. Charles Hesdorffer, a member of our hematology group, had used these retroviruses in Phase 1 human clinical trials to express a potential anti-cancer gene, the multiple drug resistance (MDR) gene.

For these studies, Dina Markowitz, a graduate student in my lab, together with Dr. Stephen Goff, a retrovirologist at Columbia, and I had developed safe and efficient defective viral cell lines that transferred and expressed human MDR genes safely in human cells. Using these lines, we achieved the expression of the MDR gene in the bone marrow of patients, but at too low levels to be of clinical significance.

In mice, we and others had shown, with γ retroviral vectors, that we could transfer and express a human β globin gene, occasionally at a high level. Dr. Harry Raftopoulos with Maureen Ward and Christine Richardson from my lab, all the way back in 1997, reported one mouse who made 20% as much human β globin as mouse β globin! However, this was a rare significant positive result in a sea of negative ones.

Efficient globin gene therapy is not reproducible using γ retroviral vectors. This is primarily because our target cells, the blood stem cells, are largely quiescent; they divide relatively infrequently. Gammaretroviruses require cell division to move their contents from the cytoplasm to the nucleus of cells. They are, therefore, relatively inefficient at having their DNA (reverse-transcribed from viral RNA) enter the nucleus of the only occasionally dividing blood stem cells.

To overcome this problem in our current clinical trial, we are using another special type of retrovirus, a so-called lentivirus, that does not require cell division for the gene-containing viral DNA to move from the cytoplasm to the nucleus of the human blood stem cells. Drs. Michel Sadelain and Philippe Leboulch have pioneered the use of lentiviruses in human globin gene therapy. Working independently, they have each shown that the anemia in mice with a disease resembling human Cooley's anemia can be alleviated significantly using lentiviruses containing a normal human β globin gene. These experiments took several years to complete and be published, and were the impetus for our human globin clinical trial in Paris.

Sadelain and Leboulch also made another critical contribution to the human β globin gene therapy field earlier. They had both previously shown that it was necessary to remove certain specific nucleotide sequences from human β globin IVS2 in order for the globin retroviral vector to function best. These studies took several years to be completed as well. Sadelain and Leboulch and their colleagues, including me, have been working on how to do efficient human β globin gene therapy in mice and humans for over 20 years.

Our magic medicine in globin gene therapy is a special kind of "vector," or carrier, to transfer our gene, the normal human β globin gene, into the patient's blood stem cells. For this purpose, we are using a slightly modified but normally functioning human β globin gene with a single base change at position 87 (β87). The human β87 globin gene leads to normally functioning human hemoglobin (Hbβ87) that at the same time is easily distinguishable from normal human hemoglobin (HbA) by laboratory testing; Hbβ87 can be quantitated by a special sensitive technique, called high pressure liquid chromatography (HPLC), that can accurately detect even small amounts of the new hemoglobin.

Using a gene different from the normal human β globin gene in the β87 globin vector (called Lentiglobin) is essential in allowing us to distinguish the hemoglobin (Hbβ87) made from our gene therapy from transfused normal hemoglobin (HbA); this is especially important in this thalassemia trial since the patients' blood will always contain lots of HbA as a result of the continued blood transfusions that patients like TK receive as part of their supportive treatment, during and after the trial.

We can detect and quantitate the amount of new added β87 globin gene in the patients' cells by using a test called polymerase chain reaction (PCR). Tests positive for the presence of the β87 gene by PCR, and for the presence of β87 globin by HPLC, are sure measures of any success we achieve in the trial. We have already obtained positive results in TK using both PCR and HPLC, although at low levels.

The first clinical trial

The defective lentiviral vector, Lentiglobin, carries our human β globin gene in our Paris trial. This vector was developed by Dr. Leboulch at MIT and has been licensed to a small biotechnology company, Genetix Pharmaceuticals Inc., which is sponsoring the trial. Dr. Ronald Dorazio is the current chief operating officer of the company, and I am the primary medical consultant to it in the US, and an equity shareholder as well. Ron and I formed the original

Genetix Pharm. in 1990, with two other scientists from Columbia University, Drs. Saul Silverstein and Stephen Goff, to do human gene therapy. The company has undergone and survived a series of mergers and private financings to remain active today.

The environment for gene therapy research has been relatively more favorable in France than it is in the US. The unexpected death of a patient in a gene therapy trial at the University of Pennsylvania in 1999, using another type of virus, adenovirus, had soured the scientific and medical community on the field in general. Before that, other approaches to gene therapy in humans had been tried in the 1970s and 1980s, with no real success.

The first largely successful human clinical gene therapy trial was done using gammaretroviruses in Paris in the late 1990s by Drs. Alain Fischer and Marina Cavazzana-Calvo and their associates. The patients in this trial were children with an immunological deficiency disease called X-SCID. This trial was widely heralded as having cured children of this otherwise fatal immune disease, results that have since been reproduced by other European investigators in London and Italy.

However, more recently, there have been major complications in some of the Paris and London X-SCID patients. After years without problems, four of the 10 apparently cured Paris patients developed acute leukemia. Since then, one patient has died, and the three other affected patients with this unexpected and unforeseen late complication are responding to chemotherapy. In the London trial, the five to ten patients treated thus far, using almost the same protocol, had no problems until very recently when one patient developed leukemia. To date, the Italian patients, with a different gene (a so-called ADA gene) being corrected, have not had any similar complications.

The unacceptable events in the French and English trials are probably related, at least in part, to the insertion of the curative gene in a blood stem cell in the vicinity of an oncogene, a gene whose expression can cause the uncontrolled growth of cells and cancer.

In X-SCID, normal T-lymphocytes lack a normal γC receptor (γC) gene, whose protein product is necessary to fight certain types of infections. In the Paris and London X-SCID trials, the corrective γC gene therapy vector contained powerful viral enhancer elements that apparently activated not just the γC gene but nearby genes as well, in a process known as insertional mutagenesis. In two of the patients who developed leukemia in the Paris trial, it has been shown that the insertion of the γC gene is near a known oncogene, LMO2. Insertional mutagenesis in a single blood stem cell in these patients resulted in a clone of cells that emerged, expanded and caused the leukemia.

The design of the Lentiglobin vector used in our Paris study avoids the potential problems of the γC vector in the X-SCID trial. First, and most importantly, viral enhancer elements are not being used to activate genes, as was the case with the γC vector. Instead, only human β globin gene-specific promoters and enhancers are used to direct human β globin gene expression, and these are only active in red blood cells. In addition, DNA sequences, called "insulators," have been added to the Lentiglobin vector to prevent the activation of oncogene sequences and the insertional mutagenesis seen in the X-SCID trial.

A brave new world

Is adding a normal gene that can insert anywhere in the chromosomes of blood stem cells the best way to do human globin gene therapy? Probably not in the long run.

Rather than the gene addition strategy we are using, ideal human globin gene therapy for Cooley's anemia as well as sickle cell anemia would be "gene correction": the correction of the single base mutation in the DNA of the mutated human thalassemia β globin gene (and the sickle cell gene as well), at its normal chromosomal position. This correction of the mutant human β globin gene can be accomplished by adding a vast excess of DNA containing the right β globin DNA sequence to correct the wrong DNA sequence (a process called homologous recombination) in the blood stem cells of patients with sickle cell disease and thalassemia. If successful, gene correction has a tremendous advantage over gene addition in that there is no possibility of insertional mutagenesis since the corrective piece of DNA used is inert, and the only change in the patient's chromosomes is that the β globin gene is corrected.

Gene correction has been successfully used to normalize gene defects, including those of a mutant human β globin gene (the sickle cell gene), in embryonic stem (ES) cells, the multipotential cells that are capable of producing any tissue of the animal. This is because, in contrast to hematopoietic (blood) stem cells (HSC), ES cells can be grown to very large numbers of cells without altering their biological properties. These large numbers of cells are required for gene therapy because gene correction is a rare event, and many individual cells must be isolated and analyzed before a rare gene-corrected cell can be obtained.

Gene correction is impractical using HSC because these blood stem cells cannot be grown to large volumes without differentiating into mature red blood cells, dying, or mutating to cause leukemia. HSC divide so poorly that obtaining large populations of these blood stem cells for this treatment is not currently feasible.

In the past two years, however, two extraordinary advances were made that have increased the possiblility that a gene correction strategy will eventually be used for gene therapy for thalassemia and sickle cell disease.

First, it has been shown by Dr. Shinya Yamanaka and his colleagues in Japan that human as well as mouse ES cells can be derived by manipulating skin cells. In these amazing experiments, the addition and expression of just four genes (Klf4, Oct4, Sox2 and c-Myc) rewire the circuitry of the differentiated skin cells which then acquire the characteristics of true ES cells. These reprogrammed cells are called induced pluripotent stem cells or iPS. Prior to this discovery, human ES cells could only be obtained from living human embryos, but this has been considered unethical by religious groups since it involves the destruction of embryos. The use of the patients' own skin cells to obtain ES cells greatly diminishes ethical concerns.

Secondly, in a paper by Drs. Rudolph Jaenisch (at MIT) and Tim Townes (at the University of Alabama), it has been shown that iPS derived from skin cells and manipulated in tissue culture can be used to cure mice with the equivalent of human sickle cell anemia. In these experiments: skin cells from these mice were converted to iPS by the addition of the above mentioned four special genes; short inert pieces of normal human β globin sequence DNA, containing sequences that correct the sickle cell mutation, were added to the skin-derived iPS of these sickle mice; the iPS were grown to large amounts and the rare iPS in which the sickle cell mutation was corrected were isolated; the globin gene-corrected iPS were then grown to large amounts, and, after being treated in cell culture with growth factors and chemicals, became hematopoietic stem cells (HSC). These β globin gene-corrected HSC, originating from the skin cells of mice, were then used to reconstitute the ablated bone marrow of the sickle mice and largely cure their sickle cell disease in a dazzling "proof-of-principle experiment."

As with addition gene therapy, there are no immunologic barriers to transplantation of the corrected cells since the skin cells originally used are those of the patient. As previously noted, a great advantage of this skin to ES to HSC gene therapy strategy over our Lentiglobin addition gene therapy approach is that it avoids the possibility of insertional mutagenesis since no new functional DNA is added to the chromosomes.

However, a potential significant danger must be addressed before the skin to ES cell approach can be used in humans; namely, that at least one of the new genes that must be added to the skin cells and expressed in these cells for their successful transformation into iPS is known to be oncogenic, and cancer has been seen frequently in mice when this technology has been used. Safer methods of generating iPS from skin cells must be developed before any human applications can be considered.

Should we wait until gene correction or another better safer strategy works before we try to cure thalassemia and sickle cell disease with our current technology? That is the current question. Presently, there are no alternatives to our Lentiglobin gene therapy approach.

Our Lentiglobin trial, in its broadest sense, is as much a bone marrow transplantation trial as a gene therapy trial. For that reason, two groups of expert investigators in Paris are involved. Dr. Eliane Gluckman, a world-renowned bone marrow transplantation expert, is one principal investigator, and TK's transplant was done under her direction at Hôpital Saint Louis in Paris.

Dr. Leboulch, a gene therapy scientist and hematologist, is the other principal investigator. He is a professor at Harvard Medical School and also the scientific director of Genetix, the sponsor of the trial. Dr. Cavazzana-Calvo, a major investigator on the X-SCID trial, with vast experience in gene transfer into blood stem cells, is in charge of the Lentiglobin gene transfer in Paris. She and her associates are at Hôpital Cochin in Paris. Dr. Yves Beuzard at Hôpital Saint Louis supervises the laboratory evaluation of the patients.

We have permission from the French governmental agencies to study up to five patients with thalassemia and five patients with sickle cell disease in this trial. A second thalassemia patient, PLB, has already been treated on the protocol, and has also recovered well from the marrow ablation. He has had a higher level of $Hb\beta87$ expression than TK and this expression has increased over time, suggesting selection for $Hb\beta87$-producing red cells.

Patients with sickle cell disease, another debilitating anemia, are also suitable candidates for this globin gene therapy trial. In β thalassemia, we only have to provide enough new human hemoglobin function for a therapeutic result, while in sickle cell disease, we have to also overcome the effects of the abnormal sickle hemoglobin that will continue to be present in the gene-corrected cells. A major reason to include patients with sickle cell disease in our current trial is that the modified human Hbβ87 hemoglobin which we are using is known to prevent sickle cell hemoglobin from causing sickling of human red cells.

In summary, after 20 years on the drawing board and with recent evidence in animal models that successful β globin gene therapy is possible, a human clinical trial in β thalassemia and sickle cell disease is in progress in Paris. People suffering from these severely debilitating diseases are being offered a new potentially curative approach to their disease in a single procedure. The benefit-risk ratio seems reasonable. The trial has begun.

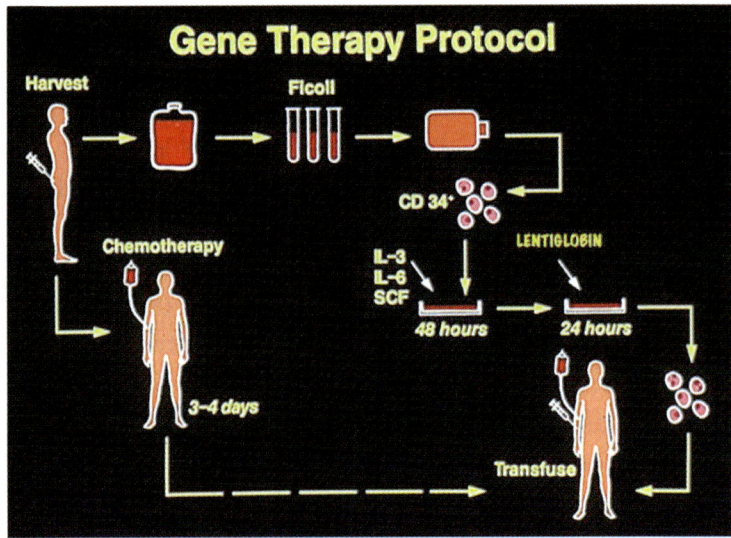

Human Lentiglobin gene therapy. Bone marrow cells are harvested. The cells are purified by Ficoll gradients and exposure to anti-CD34. The CD34+ cells are incubated with growth factors including interleukin (IL) 3, thrombopoietin (TPO), Flt3-ligand (Flt-3L) and stem cell factor (SCF) for 24–48 hrs. This cocktail has replaced IL-3, IL-6 and SCF, used previously. The cells are then exposed to Lentiglobin, the human β globin gene-containing virus, for 24 hrs. While the cells are being processed, the patient's own bone marrow is ablated using chemotherapy. Then the gene altered cells are transfused into the patient intraveneously.

Chapter 16

Genetic Roulette

In thalassemia, the inheritance of a mutant gene from each parent predictably causes "the homozygous state" and the disease. If a fetus receives a thalassemia gene from only one parent, the infant will have thalassemia trait, be "heterozygous," and be clinically well. To develop Cooley's anemia, both parents must carry the thalassemia trait gene, and transmit them to the offspring. If each parent carriers a thalassemia gene, then, statistically, one in four embryos or fetuses will inherit two mutant genes, and have the disease.

The presence of Cooley's anemia in fetuses is now detectable early in fetal life in almost all affected populations. Testing is done by analysis of fetal DNA since almost all of the single base mutations that cause thalassemia in different populations are now known and can easily be identified by clinical laboratory testing. Specific molecular probes are used to detect the specific DNA defects in any given family antenatally. Statistically, three out of four fetuses tested in families at risk will either be normal or have thalassemia trait. This testing provides the option for termination of pregnancies with affected fetuses.

This is the genetic roulette in thalassemia: the risk of serious disease remains one in four with each new pregnancy, no matter how many previous pregnancies there have been, or how many non-thalassemic or Cooley's children have resulted from these pregnancies. Even if you have three children with Cooley's anemia, a rarity, and you have a fourth pregnancy, the risk of that fourth child having Cooley's anemia is still one in four.

This uncertainty raises questions for prospective parents, both of whom have β thalassemia trait. Do they want to know about the presence of disease prior to birth? And what do they do if a fetus destined to have Cooley's anemia is diagnosed *in utero*?

When he was the President of the Cooley's Anemia Foundation in 1980, Bob Ficarra worked hard to have all Italians tested before marriage for the presence of thalassemia trait. He and others thought that education was the key to controlling or eradicating the disease. Together with Nunzio Cazzetta, and with the support of the CAF, Bob first tried to obtain legislation in New York State to establish free testing centers for thalassemia alone, but was unable to do so. He asked to have a piece of paper given to all people of Italian and Greek extraction telling them how they could obtain free screening for thalassemia trait when they appeared for pre-marital blood tests.

The only way he could accomplish his objective was to include the testing of Jews for Tay-Sachs disease and of African-Americans for sickle cell disease. This comprehensive legislation was eventually passed.

Most physicians taking care of adult patients do not consider Cooley's anemia as a diagnosis since it is primarily a pediatric disease. In addition, many Cooley's patients who live into adulthood with the disease continue to be cared for by their pediatricians; this situation keeps physicians who are caring for adults relatively unaware of this condition.

Most non-pediatricians are also insufficiently cognizant of the diagnosis of thalassemia trait. Over 99% of thalassemia carriers have small red blood cells, or low mean corpuscular volumes, or MCVs, yet obstetricians, even seeing a low MCV, do not suspect thalassemia trait as the cause. They often assume that a low MCV in a pregnant mother is due to the presence of iron deficiency, a common problem in pregnancy, and another cause of a low MCV. Thus, families without a history of children with Cooley's anemia are often under the medical radar, until they have a first child with the disease.

Testing for β thalassemia trait is not routine or easy. If there is no family history of the disease, it is especially difficult because there is usually no anemia or other tell-tale symptoms or signs in thalassemia trait carriers. They all look well. However, blood testing by machines that are routinely used to do blood counts in hospitals can strongly suggest the diagnosis of thalassemia trait from the small size of the thalassemia red cells, reflected in a low MCV.

Blood cell abnormalities such as misshapen cells seen on visual examination of the blood smear are also common, but detecting them requires staining of the blood smear (and technical expertise in identifying the abnormal cells) and this is not routinely done. Confirmation of the diagnosis of thalassemia trait then requires a special blood test which measures the HbA2 level. The amount of this minor hemoglobin component is less than 2% of the total hemoglobin in blood normally and rises to over this level (usually 4%–5%) in patients with thalassemia trait.

In the 1970s, fetal blood sampling and globin chain analysis was the only test available for the antenatal diagnosis of Cooley's anemia. At that time, the procedure carried risks and could only be done in specialized centers. By the 1980s, however, much safer antenatal testing of the DNA of fetuses at risk for thalassemia became routinely available in many centers. This testing can accurately and quickly tell if two thalassemia genes and the disease are present by analyzing cells from embryos at risk. The diagnosis can be made from the DNA of any fetal cells obtained (all fetal or adult cells have the same DNA), either early in the pregnancy by chorionic villus sampling (CVS) or later, by amniocentesis. Both of these procedures are reliable and quite safe.

Bob Ficarra, like many others in the '70s and '80s, assumed that when thalassemia families knew their genetic status through education and testing, this knowledge alone would somehow lower the incidence of Cooley's anemia. But how was that supposed to happen?

Presumably, families aware of the potential for having the trait would be screened; prospective parents in families at risk (in the US primarily those of Italian or Greek ancestry) would be tested; and then if both parents were found to be carriers, they would undergo genetic counseling to learn about the odds of having a new baby with the disease.

Potential couples in which one of the two prospective parents did not have thalassemia trait were home free; there would be no chance of having a child with Cooley's anemia. This knowledge was certainly a positive benefit of testing and relieved the minds of many in families in which there was a history of the trait or the disease.

But what about the choices of two adults who were contemplating marriage and children, each of whom was found by testing to have thalassemia trait? What were their choices? By itself, education and genetic testing of these prospective parents could not reduce the incidence of Cooley's anemia. Increased awareness of the situation does not change their risks.

One more step is needed: Action. Action that requires potential parents, both of whom carry thalassemia trait, to modify their reproductive choices. There were three alternatives in 1980, all of which still exist today.

First, the two affected potential parents could choose not to marry. They could choose alternative, thalassemia trait-free partners and all would be well medically. However, psychologically and emotionally, this would be devastating, and is generally considered an unacceptable choice.

Secondly, they could marry but not have any of their own children. They could adopt children. Again, this would solve the medical problem, but is usually unacceptable emotionally for two people in love who want and can have their own biological offspring.

The third choice for couples in which each partner is a carrier is the most common and can turn out to be the most difficult of all. Parents known to be at risk can, after genetic counseling, elect to have antenatal diagnosis performed. They can have their fetus tested for Cooley's anemia, and then decide whether to terminate the pregnancy if the diagnosis is made. Statistically, to repeat, three out of four times the fetus will be normal or only have thalassemia trait, a benign condition. And the parents are reassured.

It is the one in four fetuses destined to have Cooley's anemia that leads to necessary decision-making that in many cases is accompanied by great emotional turmoil.

Proper genetic counseling is necessary prior to antenatal testing. Prospective parents, after being informed of the odds of having a child with the disease, can either take their chances, or have their fetuses tested. As indicated earlier, this testing can easily be done today, as early as 11 weeks of gestation, by examining the DNA of fetal cells by chorionic villus sampling (CVS), or later in pregnancy by amniocentesis.

The obvious advantage of CVS is that it can be done when the fetus is undetectable externally. By the time of amniocentesis, the pregnancy is usually obvious and the baby might be kicking. The earlier the diagnostic test is done, the less the emotional toll it takes on those parents who have to make a decision about what to do next.

In one large study, in which I participated, of antenatal diagnosis in pregnancies in which both parents had sickle cell trait and were at risk for having a fetus with severe sickle disease, the parents were more likely to terminate affected pregnancies after CVS, done earlier, than after amniocentesis, done later in the pregnancy.

For two parents with thalassemia trait the question of what they will do if the embryo or fetus is destined to have Cooley's anemia has enormously complex implications. Termination of pregnancy is the only way to prevent the disease, and prospective parents at that point have the option of making the choice. However, the ethical, cultural and medical dilemmas surrounding this decision can be overwhelming. This is especially true as the treatment of the disease and patients' life expectancy and quality of life improve.

The Cooley's anemia community in most developed countries is religious. Italians in the US, the primary subjects in this book, are predominantly Catholic. The Catholic Church in New York and worldwide has never accepted therapeutic abortion as an option in preventing thalassemia.

Bob Ficarra tells the story of one of his encounters with a Catholic Church official regarding antenatal diagnosis. He suggested that a piece of paper be given to couples of Italian extraction, informing them of free thalassemia trait testing, when they appeared for a "pre-Cana conference" with a priest before marriage (a Catholic ritual).

Ficarra: "I'd like your blessing to make people aware of a program of testing Italians at risk for thalassemia. One out of four of their children will have a potentially fatal disease that we can prevent."

Church: "What would happen next?"

Ficarra: "If the two parents both have the trait, then the fetus would be tested for the presence of the disease."

Church: "What would happen next?"

Ficarra: "If the fetus is affected with the disease, termination of pregnancy for medical indications would be considered."

Church: "That is not acceptable. Why do the testing at all?"

In several European communities with very high rates of thalassemia - in Sardinia; in Cyprus; and on the mainland of Greece - the incidence of Cooley's anemia has been dramatically reduced since the 1980s when thalassemia testing, genetic counseling, and antenatal diagnosis became available.

There are many stories told of how this might have happened. The Catholic Church and the Greek Orthodox Church throughout the world have consistently been against abortion, even in cases where Cooley's anemia is the known consequence of inaction. However, it is also known that in these communities, premarital genetic testing was instituted widely at the urging of physicians and families because of the many patients with severe disease, and presumably this testing was supported (or rather not opposed) by the religious community.

What happened next is still not completely clear. Was the choice of abandoning a prospective mate for an unaffected partner a more viable option than expected? Were there less pregnancies in families at risk? Was there some acceptance of antenatal testing and pregnancy termination? Who knows?

Dr. Antonio Cao, a major figure in the control and treatment of Cooley's anemia in Sardinia, published the following in 1987 and 1990:

In 1987: "The characteristics and the effectiveness of programmes designed to prevent β-thalassemia major present in high frequency in several areas of the world such as Cyprus, Greece and Sardinia are reviewed. All these programmes are based on heterozygote detection, counselling and foetal diagnosis. The target population for screening have been couples at marriage, conception, or early pregnancy. Awareness of the problem and involvement of the population was achieved via mass-media or personal approaches through lectures or discussions.

"Parent's Associations were consulted and have been actively involved. Information leaflets have been made available to prospective couples at several critical areas. Education on thalassemias was introduced into the school curriculum. Counselling was based on a private interview at which the several options available were discussed with the individual carrier or the couple.

"Prenatal diagnosis was chosen by the large majority of couples counselled. All these programmes resulted in a decline of thalassemia major births by 50%–97%. The reasons for residual cases were mostly lack of information and, less frequently, misdiagnoses or refusal of foetal diagnoses."

In 1990, Dr. Cao continued: "Prenatal diagnosis was initially carried out by fetal blood analysis; since 1983, it has been done by DNA analysis on non-amplified or amplified DNA. Different chorionic villous sampling procedures have been used. Nowadays, we have adopted the transabdominal approach because, in our experience, it seems to be associated with a low risk (2%) of fetal mortality.

" ...On the whole we have so far carried out 2711 prenatal tests: 1130 by fetal blood analysis, 1156 by oligonucleotide hybridization on electrophoretically separated DNA fragments, and 425 by dot-blot analysis on amplified DNA with allele-specific oligonucleotide probes. Two errors occurred by fetal blood analysis and none by DNA analysis. The incidence of thalassemia major declined from 1:250 live births in the absence of prevention to 1:1000 after the establishment of this program, indicating that carrier screening and prenatal diagnosis are effective means for preventing thalassemia major at the population level."

Dr. Cao told me in 2008, "At the moment the number of new cases per year is in fact four to five instead of the 120 we saw before starting the program."

Precisely how religious leaders interacted with the medical community in Sardinia, and in Greece and Cyprus, regarding the termination of pregnancies of fetuses diagnosed with the disease, is unknown. There was apparently a tacit understanding with the religious community that allowed for the termination of pregnancy in cases where the severe disease was diagnosed in the fetus.

Some countries have more openly receptive attitudes towards antenatal diagnosis and abortion. In India, Israel and Gaza, small studies have reported that all affected families opted for termination of pregnancy. How fast or slowly acceptance of new technology and genetic counseling moves in different countries obviously depends to a great extent on their unique cultural and religious heritages.

Another brave new world

More recently, a more complicated procedure mentioned earlier, *in vitro* fertilization (IVF) and antenatal selection of embryos, commonly used to help couples with fertility problems, constitutes a new option that is available for the antenatal management of Cooley's anemia and other diseases. In this procedure, eggs are harvested from the prospective mother and fertilized by sperm from the prospective father in a culture dish incubated outside the body. The fertilized eggs thus obtained are then grown to the four-cell stage and genetically tested.

At the four-cell embryonic stage, it is possible to tease away one of the four cells to test for the presence of thalassemia genes. If chosen for implantation, the three remaining cells, capable of developing into a normal fetus, can then be grown further in the dish and implanted into the mother or surrogate. For the antenatal diagnosis of thalassemia, this procedure can be used to determine whether a potential fetus is thalassemia-free, has thalassemia trait, or is destined to have Cooley's anemia. The goal of the preimplantation diagnosis is to ensure that no fetuses with Cooley's anemia are implanted.

This is essentially the same procedure that is currently being used to ensure the birth of HLA-identical fetuses who can be donors for siblings requiring ABMT: siblings with thalassemia or sickle cell disease, or those with leukemia or other cancers. In this approach to thalassemia, only embryos that are either thalassemia-free or have thalassemia trait are chosen for growth and implantation. The embryos with the disease are not used. This procedure could be more acceptable to prospective parents and those with objections to abortion since it is done before pregnancy is initiated *in utero* and unused embryos can simply be frozen away.

In vitro fertilization and embryo testing thus provide another way to address the medical and ethical issues of antenatal diagnosis and termination of pregnancy with fetuses at risk for Cooley's anemia. However, it is also an extremely expensive and only moderately successful approach to a normal full-term pregnancy. For most families, the prohibitive cost of *in vitro* fertilization and preimplantation testing makes it an unlikely option at present.

To repeat: The reproductive decisions of prospective parents at risk for having a child with Cooley's anemia, with all the facts and options at hand, are complicated. Antenatal diagnosis and termination of pregnancy or *in vitro* fertilization approaches are available. The decisions are further complicated by the fact that currently patients with the disease are living healthy lives with adherence to a program of transfusions and Desferal therapy. Oral iron chelators are available and may well replace the intensive needle-sticking regimen that Desferal requires. Bone marrow transplantation of patients with compatible donors is a potential totally curative option for some patients. And there are clearly different religious, social and cultural attitudes that affect reproductive choices in families at risk for having children with Cooley's anemia.

The question of whether to do antenatal diagnosis often occurs in families that already have a child with the disease. Parents ask themselves what do we do, if anything, about the next pregnancy? The decision regarding antenatal diagnosis for a second child, after having a first child with thalassemia, has deep ramifications. When I asked one father of a child with Cooley's anemia in the US about family planning, and the potential use of antenatal diagnosis in determining the outcome of a future pregnancy, he said, " How could I ever look my first child in the eye in the same way again if I took the life of someone just like her?

"It flies in the face of the meaning of parenting. If God chooses me to have and raise a baby, any baby, I take it as a privilege. God works that way. God does not pick wrong children. There are no wrong children."

Another father of a Cooley's anemia patient told me the same thing: "My [thalassemia] child has a good life and we could have had another child with the disease. It would have been an offense to my child with the disease if we would not have accepted a baby with Cooley's anemia, if that's what the new baby had."

Chapter 17

Modern Times

The Chiecos

Michelle Chieco is a sophomore at Fairfield College with Cooley's anemia who comes to White Plains Hospital every three weeks to be transfused, as she has for the past 18 years, from the age of one. The second of three daughters of Peter and Rose Ann Chieco, she is the only one in the family with the disease. From the time her illness was discovered, her parents have been totally devoted to her optimal care. They have ensured that she receives her transfusions and maintains a hemoglobin of 10.

Peter says she's "pretty much a normal college kid." She is studying to be a nurse. When she was a child, Peter usually mixed her Desferal solution every night and set up the pump used to administer the drug.

Michelle has had Desferal injected under the skin (subcutaneously) of her thighs since she was four years old. "It is the only place she wants it to be infused so that needles do not violate any other part of her body, not her abdomen, her arms or other parts of her legs," Peter says.

Her ferritin, a measure of the iron load in her body, is at a reasonable level for a thalassemia patient being transfused, and she has had essentially normal growth and development. Michelle is an example of the success of the current comprehensive treatment of

Cooley's anemia as it exists today. Provide enough blood by transfusion, get rid of the excess iron with Desferal, be compliant and survive.

I ask Peter about new treatments. "Desferal works," he says. "Bone marrow transplant is scary. She could die. Gene therapy is a long way off. You guys [meaning, us gene therapists] seem to climb one mountain and get to the top and then there's a valley and another mountain ahead before you can reach your goal. It all takes time. I don't know when you'll reach your goal."

But Peter has never been satisfied with the reality of his child sticking herself with a needle almost every night of her life, and having her sleep with a pump attached to a syringe taped to her thigh to deliver her magic therapy for 8–12 hours overnight. It is a trial of compliance and patience that is both demanding and painful.

Peter has long been an advocate for the availability to Michelle, and other patients, of an oral iron chelator to replace Desferal and its injections. He has intensively championed this cause for 15 years, and finally Exjade, such a drug, has been approved in the US and is available.

Ever since another oral iron chelator, L1, was developed in England in the mid 1980s, Peter has led efforts by the CAF to have it approved for use in the US. He has tried to convince the scientific and medical community and politicians in Washington to do what it takes to hasten FDA approval for this drug, which is much cheaper than Exjade, and has been used extensively and found to be effective in Europe and Asia. The CAF has had a consistent single creed in the 43 years I have been associated with it that I can vouch for: do whatever you can to help and cure patients with Cooley's disease. And Peter's aggressive attitude in the struggle to get L1 approved has simply been a continued adherence to that creed.

But a variety of unfortunate circumstances in the development of L1, discussed earlier, has prevented the approval of this drug by the FDA. "I went to everyone in Washington and to the clinicians in the US, but couldn't get them to push for FDA approval of L1. I told them L1 is approved in Europe and in many other countries, and it has been shown to work, but I could not convince those who make these decisions to do so. Patients in the US died because of that. The experts said 'No. We have to have Phase III trials, proof that it works.'" I have apologized to Peter for having been among those experts; I believed Nancy Olivieri's reports about the liver toxicity of L1 at that time.

"I told them all that Desferal works, if you can have your child use it the way Michelle uses it," Peter recalls, "but there are children and families that cannot deal with mixing solutions every night and giving injections to their little chidren. And as they grow up, the children find it harder to comply. I tried to tell these people that we would have saved children's lives, in the more than 10 years before the availability of Exjade.

"If they couldn't tolerate Desferal for any reason, they should have been allowed to take L1. We could have saved a lot of children's lives by allowing its use in the US as in Europe. It has proven to be an effective drug."

Peter Chieco was born in New Rochelle on August 18, 1960. Three of his grandparents and his parents, Leonard and Marie, were born in the US although they all had extended families in Italy: in Bari, Calabria, and Naples. His grandfather, Peter, whose name he carries, came to the US in 1936.

Peter was the middle child. He has a sister two years older and a brother two years younger. They both were diagnosed with mild forms of cystic fibrosis, another genetic disorder, but Peter has been well all of his life. I met him 20 years ago. He had an emotional maturity and intensity beyond his years and I was surprised to learn that he was only two years older than my oldest child. I have always loved Peter Chieco as a person. We spoke in his Greenwich, Connecticut office.

Peter worked as a child and young adult in the family business, Anna's Harbor restaurant, now gone, a wonderful expansive seafood place on the water in City Island, New York. My wife and I used to eat there regularly, not knowing the Chieco family owned it.

Peter went to a Catholic school, Blessed Sacrament, and then to Fordham University, where he earned a degree in finance in 1982. He has worked in financial services since graduation, and is currently a successful Smith Barney financial adviser working with stock trades, estate planning, and wealth management.

If I weren't a TIAA-CREF person, and had any money to invest, I'd want someone like Peter to be my financial advisor. He's straight as an arrow, clear as one can be in his thinking and logic, and a joy to be with. He has a great sense of humor, and is a handsome devil. I couldn't think of a better friend, even though we are so far apart in age.

Peter married Rose Ann Fantino in 1985. Her background was all Italian as well, Southern Italian. Peter and Rose Ann went for "blood tests," and were not told that anything was wrong with them. No one in the family had ever heard of Cooley's anemia or thalassemia until Michelle was diagnosed with the disease.

But Peter had heard from his grandfather, Peter, and his father, Leonard, that they were "anemic" in an unspecified way, and that he was too. Rose Ann, his wife, was told by the family doctor that she was fine, not anemic.

Deanna, the Chieco's first child was born on September 7, 1987. She was a fine healthy baby, dark and beautiful, who was walking at nine months. Michelle was born less than a year later, on July 28, 1988. "She was born by breach and was blonde and light in complexion with chubby cheeks. She looked well," according to Peter.

At four to five months of age, Rose Ann's mother, Jennie, was worried that Michelle "was cranky and pale." The Chiecos were living with Jennie in her big empty house since her husband had recently died. "Michelle often put her head down on the ground and lay on her side as if she was tired and had no energy. She was also a picky eater. And unlike Deanna, she didn't seem to have the strength to walk, even at close to a year," Peter recalls.

Rose Ann had been taking the children to the family doctor every few months for check-ups. When Michelle was 11 months of age, her hemoglobin was found to be seven grams percent, quite low. The doctor told the Chiecos "she's anemic. It must be an infection. I'm putting her in the hospital."

At that time, the family doctor consulted a hematologist, Dr. Judith Marcus, who looked over all the blood tests that had been done on the Chiecos and their children over the years. She said almost immediately: "Michelle has Cooley's anemia. That's what it's got to be."

Among the pile of lab results on the family, Dr. Marcus found that Peter and Rose Ann had very low MCVs, the specific blood value that is almost always abnormal in individuals with thalassemia trait, the carrier state of thalassemia. She immediately surmised that Michelle had Cooley's anemia, the severe form of the disease. Peter also discovered then that his grandfather Peter and his father also had the same low MCVs, the cause of their "anemia."

All of the many low MCVs were recorded in the Chiecos' charts, including those from the Chiecos' pre-marital testing. The oversights, and lack of recognition of the significance of these values in so many family members, shocked Peter and Rose Ann. They could have known before Rose Ann's pregnancies that they were facing the possibility of having a child with Cooley's anemia.

"It may not have altered how we would have approached the situation, but we were really angry and frustrated that the numbers were there all the time for everyone to see, and no one saw them until it was almost too late, and no one did anything about it."

It was late at night when Dr. Marcus told the Chiecos Michelle's diagnosis. The next morning, their family doctor told them, "She has a fatal disease." That's a sentence that I've never used in over 40 years of medical practice, and I don't think should be used by physicians. All of us, in a sense, have fatal diseases: it's simply when we get them and when they kill us that varies, but we will all die of something.

The next day, Dr. Marcus returned and told them that Cooley's anemia was "serious but treatable," and immediately referred them to the Cooley's Anemia Foundation for further guidance and advice.

"From that day on," Peter says, "The Cooley's Anemia Foundation was my therapy." At the first meeting he attended shortly therafter, he saw the patients "with deformed faces and gray skin." He and Rose Ann were shocked and depressed.

At the end of the meeting, Dr. Sergio Piomelli, who was chairman of the CAF Medical Board at that time, told them "Don't worry. Your children are going to do well. We have better ways of treating patients today than some of what you see here." His words were particularly encouraging and reassuring. Peter remembers them to this day.

One week later, the CAF called Peter and asked him if he would like to be the new President of the Foundation. He had so impressed people with his sensitivity, devotion, and intensity at that first meeting. He declined, but joined the Executive Board then, and became President later. "My therapy after Michelle's diagnosis was getting involved with the CAF. It saved me and drove me."

The Chiecos decided to begin Michelle's transfusions almost immediately after her diagnosis was made, despite advice from some hematologists who encouraged them to "wait and see" whether she had a milder form of the disease, thalassemia intermedia. This condition, as I've discussed earlier in the book, is one in which patients stabilize their hemoglobins in the seven to nine range, and do not require regular blood transfusions to survive, as Cooley's anemia patients do. In most cases, it becomes obvious soon enough on the basis of whether or not patients maintain their hemoglobin level without transfusions as to whether a given patient either has thalassemia intermedia or full-blown Cooley's anemia.

The day before her birthday, while wrapped in a blanket and being held still by a nurse, and with her parents watching, Michelle received her first transfusion into a vein in her arm. The next day, her birthday, Michelle began walking. It was a miraculous transformation. Instead of being pale and tired, Michelle was almost instantly rosy-cheeked and smiling and happy and walking. "We cried with joy that day," Peter recalls.

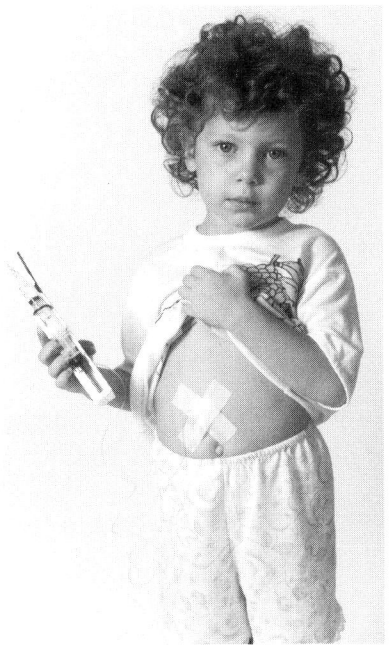

Michelle Chieco at three years old, holding her Desferal pump (courtesy of CAF).

Since then, Michelle has been transfused with blood donated by family and friends of the Chiecos to the White Plains Hospital, specifically for her use every three weeks. To this day.

Michelle began receiving Desferal at age two. It is the practice to delay iron chelation until the blood ferritin level, a measure of the total iron in the body, is elevated. This is important since iron has many important biologic functions in all of us, especially in the infant and young child, and there can be side effects if too much iron is removed when the iron stored in the cells of the body is not adequate.

Michelle appeared to be just another typical student at school, except for her absences for transfusions. Until seventh grade, she hadn't told anyone at school of her illness. Then she confided to a boy that she had Cooley's anemia, and he was quoted as saying, "I used to like her but not anymore. I could get a disease from her that could make me sick." When word spread through the school to her girl friends, Michelle's relationships with them were also never quite the same. She transferred to Sacred Heart in Greenwich the next year.

By age 14, Michelle was traveling with Peter extensively, talking in front of different groups, increasing awareness of Cooley's anemia and raising money for the CAF. Her self-confidence had increased as a result of her public appearances, "and so did her self-esteem," according to Peter.

Despite her earlier school unpleasantries, Michelle's experience as a high school student at Sacred Heart, an all-girls school, was great. The fact that she had Cooley's anemia, however, was not widely known to her school-mates until she went with her class on a field trip to Washington when she was 16.

During the field trip, Michelle left the group and joined Peter, and testified before the House Appropriations Committee, which was having hearings about federal funding for genetic disorders. When Peter and Michelle arrived in the committee room, it was packed with Congressmen and spectators, and they soon found out why: Julia Roberts, the actress, was scheduled to testify before Michelle about another genetic disease, Rett Syndrome, and everyone was there to see and hear Julia Roberts.

After her testimony, the actress went up to the official area, usually reserved for congressmen and congresswomen only, for hugs and congratulations. Michelle said to Peter, "Why are those men acting so silly?" Peter told her that was because "men act like fools when women are powerful and pretty."

When Julia Roberts left and it was Michelle's turn to testify, "about three-quarters of the audience and congressional officials left the room," Peter recalls. "Then Michelle gave the best presentation I've ever heard, describing the inheritance of Cooley's anemia. She brought along four apples: one bright red, two reddish-green, and one green. She explained to them that in families with this disease, one out of four babies born is a red apple, perfectly well. Two are reddish-green apples: they carry the trait but are well also; and one is a green apple with the disease.

"I am the green apple," she said. And then she described her life and what she needed to do to stay healthy. After the presentation, Peter took her back to join up with her class touring Washington.

When the class returned to Greenwich, the teacher asked Michelle to tell the class what had happened in her congressional appearance in Washington. She told them the whole truth, including the fact that she was the green apple. The response of the class was very supportive and gratifying. Not only were there hugs and kisses and support from her classmates, they even insisted on sponsoring a fund-raiser for the Cooley's Foundation. "Michelle kept pressing me to meet with a classmate of hers who wanted to organize the fundraiser, and I kept putting her off." He finally relented, and was glad he did.

"This 16-year old was more mature and understanding than most adults. She had all the details worked out. We had a great

Michelle and Peter Chieco (courtesy of CAF).

fund-raiser, despite my qualms about involving Michelle's class-mates based on our previous bad experience."

In 1994, Peter and Rose Ann decided to have another child. "Michelle has a good life and we were willing to have another child with the disease if that's what happened. We also had a one in four chance of having a potential bone marrow donor for Michelle in the new baby," Peter recalls.

The only thing that the Chiecos used antenatal diagnosis for was to prepare themselves as early in the pregnancy as possible for the eventual outcome. To do this, they had CVS analysis of the baby's DNA done at week 12. Jennie Rose was born in 1995 with thalassemia trait. The only disappointment that Peter felt was that the Chieco name will not be passed on. His siblings have no children.

He is very grateful for how life has gone for his three girls so far, and has a good shot at another special prize, being a grandfather someday.

Chapter 18

Love Unbound

Frank Somma, the President of the Cooley's Anemia Foundation, as I write this, is a man full of compassion, love and understanding. He has written many books emphasizing empathy, passion and commitment as the steps to success in life. In his book, "Frank Somma: Weekly Thought 2007," he says, "Remember, it's not what's going on around you, or to you, but rather what is going on inside of you that makes the difference." And he lives the creed, as best as he can. He wants all of us to see the brighter sides of our lives. Wallowing in misfortune is not helpful.

Frank's strong religious beliefs guide his attitudes toward family and life in general. He has two children. The first, Christine, was born healthy. Then, four years later, he had Alicia, who has Cooley's anemia. At that time, he and Deborah, his wife, had wanted five children. But Frank knew immediately that he would never have another child when, one day, he saw little Alicia, who was less than a year old, being stuck with a needle for the fourth time by a technician trying to draw her blood for a blood test prior to a transfusion. She was howling. On that day, he knew he could never subject another child of his own to that ordeal.

Frank is the most recent of several extraordinary people who have been the President of the CAF whom I have known well. All have been truly committed, dedicated, strong and unflinching people with a single goal in mind: making the lives of patients and their families better, whatever it takes. I don't believe any of them care what the specific cure for the disease will be. Their only concern is that life will be better for patients with the disease as a result of their efforts, and, of course, the sooner the better.

Frank Somma was born in Staten Island, New York, on July 30, 1958, a third generation American with grandparents of Irish-German heritage on one side, and Italian on the other. His father always told him, "You have Mediterranean anemia." This was probably on the basis of a low MCV. Frank knew that he had inherited thalassemia trait from his father, but he had no idea that meant that he might have a child with Cooley's anemia.

Frank served in the Navy Presidential Honor Guard from 1978 to 1982. He married his love, Deborah, in 1980. For the past 23 years, Frank has been a highly successful salesman and executive for Candle Business Systems Inc., one of the largest office equipment and supply dealers in the New York–New Jersey area. We talked in one of Frank's many offices.

Alicia was born on March 20, 1985. Christine, her older sister, born in 1981, had been pampered by her parents as an infant. She was never allowed to cry for more than a few minutes. Frank remembers that he and Deborah were cautioned by everyone not to repeat their hypersensitivity to an infant's crying with Alicia.

"Let her cry," they were told. And they did, Frank recalls with regret.

"If we had known that she was anemic and sick, we certainly wouldn't have let her cry for a minute."

At eight months of age, Alicia was still crying a lot. She "looked orange" and was finally diagnosed with Cooley's anemia. The hematologist who first informed Frank and Deborah of the prognosis for Alicia came close to emotionally ravaging them. He told them, "She will not keep up with other children physically or in school. She will be disfigured facially. She will likely die before she is 16."

The only positive act of this hematologist was that he referred the Sommas to Dr. Patricia Giardina at New York Hospital that same day. Alicia and her parents saw Dr. Giardina the very next morning at 8 AM. "She saved our lives. And has been there for us ever since," says Frank.

"It was a miracle. She told us immediately that Alicia would not be disfigured or delayed in her development; that there was no predictable life expectancy for her."

"Best of all," Frank recalls, "from what Dr. Giardina told us, I thought that Alicia would be a grandmother some day."

Alicia began her transfusions that same day at New York Hospital. She was transfused to a hemoglobin level of above 10 from a level of six and her health improved immediately.

She was given Desferal subcutaneously beginning at age two, and received it seven nights a week for many years. More recently, she has taken Desferal five times a week. She had a short break in her compliance with Desferal as a teenager, but is now back administering it to herself on the required rigid schedule to maintain negative iron balance.

Alicia's physical development, including her female sexual maturation, often affected in thalassemia patients, has been normal. This may have been aided by the human growth hormone injections she received every day from the age of nine to 13. Alicia is currently continuing her education and is working for the CAF as a fundraiser. As often as she can, she uses the Desferal pump from 9 AM to 6 PM, freeing herself from its hold so that she can do things like any other young person at night. She is transfused every three to four weeks.

The Sommas have been very open about Alicia's illness, and have been important fund-raisers for the CAF. Alicia has been a spokesperson for the Foundation as well, and has also testified before Congress on behalf of funding for the disease.

Frank thinks that, "Adversity can build character. I think Alicia's illness has made her emotionally stronger than she would have been without it."

My opinion is not that different from Frank's. It takes great emotional strength to survive as well as Alicia has with Cooley's anemia. Not only does it require super-parental guidance and excellent medical care, it requires a special kind of person as well, someone like Alicia and Michelle, or Linda D. and Amy P., who have been discussed earlier. They are people who realize what has to be done to stay alive with Cooley's anemia and do what has to be done.

As Dr. Alan Cohen says, "We are lucky to have the tools to manage both the anemia and the iron overload." One cannot overstate the importance of being treated at an excellent facility by excellent doctors, and being cared for by a loving family. Surviving Cooley's anemia is no different than surviving other severe illnesses; they all require personal fortitude and constant vigilance in adhering to treatment. However, these characteristics are certainly not the only determinants of disease outcomes. Surely, the vagaries of biology and luck play major roles in this process as well, but, in addition, personal fortitude and commitment to treatment make a difference.

As always, the Cooley's Anemia Foundation is most interested in finding a cure for the disease. And Frank and Gina Cioffi, the current director of the CAF, have targeted globin gene therapy as an important research area in their "search for a cure."

As far as the use of oral iron chelators is concerned, I was surprised to hear from Frank that Alicia was not yet switching to Exjade, the recently FDA-approved oral iron chelator, in order to avoid the needle sticks and discomfort of Desferal. "She feels she is doing extremely well with transfusions and Desferal, and she doesn't want to rock the boat. She hears there are problems with Exjade. Maybe she'll try L1 [the other oral iron chelator in use] when it becomes available in the US."

Frank says that the availability of oral iron chelators, like Exjade, in the US has not solved the problem of compliance with the treatment necessary to control iron overload in many Cooley's patients. "There's still a problem with compliance, even with oral treatment, in many families."

Compliance is a problem that progress in clinical care has obviously not eliminated, and Frank is right to be concerned about it as an ongoing issue. The dynamics of noncompliance are not only complex, but vary from family to family, individual to individual. Psychosocial therapeutic support services for an individual or family may be an additional treatment option that is needed to more effectively address this critical issue in the care of patients with Cooley's anemia.

Frank, a tiger at pushing the CAF's goals, is also finding it difficult today to cultivate a group of younger volunteers for future leadership roles in the Cooley's Anemia Foundation, as was the norm in the past. Frank thinks that this is due to a general change in people's values in the last two decades. "People used to have Sundays and dinner time with the family, and family and community were important. Now it's television, the internet, money, and business 24/7. There's no time to go to meetings, work for other people, or talk to people anymore.

"The continued dedication of old-timers like Connie Paradiso and Nunzio Cazzetta and Bob Ficarra, the giants in the CAF, is amazing to me. They continue to work in fund-raising, long after their family members with the disease are gone."

This generational change may also be due to the fact that younger parents are suffering less because of better standardized treatment and so they and their families are less involved with support groups *per se*. With things going well for their children, they do not feel the need to be as involved with groups like the CAF as much as the older generation of parents did. They just give money.

Whatever the reasons it is happening, Frank is concerned with the decrease in personal involvement of these younger parents in the Cooley's anemia community in the US, even as awareness of, and fund-raising for, genetic diseases are increasing in the public

Alicia and Frank Somma (courtesy of CAF).

domain. Some of the roles that parents previously played in the process of advocating for the support of research and clinical care are being filled by the patients themselves: in the US, by the Thalassemia Action Group, TAG, and by international groups such as the Thalassemia International Foundation (TIF) and the World Health Organization (WHO).

Chapter 19

The Last Chapter

While this is the end of the book, it is certainly not the last chapter in the story of Cooley's anemia. Gina Cioffi is the current Executive Director of the Cooley's Anemia Foundation. She manages all of its activities: research, public relations, patient care, outreach, and fund-raising. Essentially, Gina is a patient advocate, a director of activities, a politician, and a great friend to all of those involved with Cooley's anemia.

Gina has been an extremely empathetic person all her life. As a child growing up in Long Branch, New Jersey, she was always drawn to the problems of sick patients. She remembers raising money for muscular dystrophy as a child, and participating in telethons for these patients. She became focused on Cooley's anemia in the 1970s when she was in middle school and she had a guidance counselor (Richard Counte) with the disease. "He was the little green guy, martian-colored green from his iron overload, who struggled to walk down the hall.

"He was a little guy who had one of the first Desferal pumps. His middle protruded because of the pump wrapped around his belly. He was a vision of courage. I still think of him."

Gina's father was the mayor of Long Branch, and she was always very aware of causes that involved being in the service of others for the public good. Her mother, Jean De Stefano, was a role model for Gina who encouraged her interest in charitable

Gina Cioffi, Executive Director of the Cooley's Anemia Foundation
(courtesy of CAF).

causes, and the underprivileged and the underserved. Public service excited Gina as she was growing up and continues to do so today.

Gina talks about her "heroes" in Cooley's anemia. "Of course, there are the patients and families first. They suffer with the disease, and they try to survive.

"Then there are the doctors who take care of our patients. People like Alan Cohen, Pat Giardina, Sergio Piomelli, and Elliot Vichinsky. They save lives.

"Then there are all the volunteers and organizations that raise money for the patients and families, like the Cooley's Anemia Foundation and the Sons of Italy and Ahepa, all dedicated to care, research, education, and outreach.

"The Federal government has also done so much to provide money for Cooley's anemia research and, more recently, clinical care, establishing thalassemia treatment centers nationwide in the

Thalassemia Clinical Research Network. These have been long in coming, but are most welcome.

"NIH people like David Badman and Alan Levine worked hard, in the '70s through the '90s, to foster, defend, and supervise large research grants for Cooley's anemia.

"And lobbying efforts in Washington for the disease, supported by the Foundation, have led to new programs being funded by Congress. The efforts of John Grupenhoff earlier, and Lyle Dennis more recently, have been very fruitful in working with Congressional committees to support millions of dollars in NIH and other grants benefiting thalassemia.

"Then, of course, there are the researchers like you, who have pushed the field forward and made thalassemia the poster child for genetic diseases.

"I'm also very proud of the Cooley's Anemia Foundation. For over 50 years, it has been a stable provider of services to patients with the disease. And many of the people have been around continuously over this time. People like Nunzio Cazzetta and Connie Paradiso, and the Ficarra family. There has been stability. Even our office staff has been dedicated and long-lasting."

Gina went to St. Mary's College in California and then to Catholic University Columbus School of Law. She was always interested in "social justice, environmental issues, non-profit organizations, working for the public good." She had a job in Congress while going to law school at night. She worked for Representative Frank Pallone, who became the chairman of the House Subcommittee on Health. He was from Long Branch and was a prominent pro-environmental Democrat.

She was working on environmental issues, including solid waste transport, when she received an offer to become the full-time Executive Director of the CAF in 1994. She took it, and was an almost immediate success in the job. I remember the change in the management of the Foundation with her leadership soon after she arrived and since then. She was organized and focused in her goals.

She remembers, "I fell in love with the patients. Everyone who works here feels the same way, and the Board is so devoted to working on the problems and so caring. Everyone works together. It has been a marvelous experience."

Gina was a strong advocate of supporting the program of Cooley's Anemia fellowships, a program which I have fostered and participated in since the 1960s. This program provides funding for young investigators doing research (including myself back then), has expanded over the past two decades, and has done well. The CAF and Gina have ensured that adequate resources are provided by the Foundation to make the fellowship program attractive to young scientists in the thalassemia community, not only in the US, but around the world.

And she is active in planning the Cooley's Anemia Symposia (the last one was held in 2005, the next one is scheduled for 2009) which are supported by the Foundation and the NIH. These symposia bring together researchers, clinicians, and patients and their families from around the world.

By 2000, Gina had the Foundation on a firm organizational footing, and she decided to apply her skills to other positions in the healthcare public relations sector, with a focus on patient advocacy. She worked for two large public relations firms in consecutive high level jobs, but was "really unhappy with both of them," and resigned after six months at each.

As she was leaving her second PR job, she received a call from Apotex, the pharmaceutical firm in Canada that was beginning its first new drug outreach program for L1, an oral iron chelator for thalassemia. She was excited to be involved in the management of patient relations for this long-awaited drug and took a job there. She was instrumental in getting the story about L1 out to the public and to patients, and in promoting its use.

As I have said earlier, L1 is widely used in Europe as a treatment for iron overload in thalassemia, but is still not approved by the FDA for use in the US.

In 2005, Frank Somma, then President of the CAF, called Gina and asked her to return to the Foundation as Executive Director. She was ready and happy to do so. There were many new well-supported activities at the Foundation, and a new vibrancy brought by Frank which continues to the present time.

Gina says: "There is new support by the Centers for Disease Control and Prevention (CDC) in blood safety, and also by the NIH for the new Thalassemia Clinical Research Network, and for new clinical approaches and protocols to improve patient care. There is so much of the new: new strategies for raising money and finding cures; new options available for care, such as the new oral-iron chelators and bone marrow transplantation; new approaches to antenatal diagnosis; and new potential therapeutic interventions.

"And there are also new patients. We now have many immigrants to the United States from around the world, with new and different forms of thalassemia. These patients are from Pakistan, China, Southeast Asia, and Eastern Europe. They need special expertise. Many have forms of thalassemia that may or may not require transfusions. We are trying to provide these new additions to our thalassemia family with the best advice and care we can."

Gina Cioffi has been a model caregiver in the grandest way for patients with thalassemia. She is a special person, one of the many it has been my pleasure to work with in my years with the Foundation. Gina Cioffi and her associates have been, and continue to be, for me, the best of the best.

Alan Cohen characterizes the Cooley's Anemia Foundation as "the mouse that roared." "They really work at raising money, but are only a minor contributor to the total amount of thalassemia research dollars, compared to the NIH. Yet, they continually push new incentives in the disease, and have consistently optimized the use of their funds for research, education and patient care," he says.

I am particularly grateful to the CAF for supporting me early in my career, and also for supporting research fellows in my laboratory later on. I have also, over the past five decades, been as close to the people in the Cooley's Anemia Foundation as I have to any group outside of my own family. In the days when I was their medical director, I remember the intensity and good feelings at the monthly meetings of the Cooley's Executive Committee.

I am also extremely grateful to the Federal government and the NIH for their generous support of my research in Cooley's anemia over the years.

Alan Cohen says, " There are only two things that I worry about in the future of thalassemia. One is that the healthcare system in the US will allow the optimal care of Cooley's anemia patients to continue in ways that will satisfy the Foundation, and the other is that younger people will come forward to replace me and other clinicians currently treating the disease who are getting older."

He adds, "When I was medical director of the Foundation [in the 1990's] I remember missing my train at Penn Station once because the meeting of the Cooley's Executive Committee went late and I ended up on the sleeper train, arriving in Philadelphia at 5 AM with very little sleep despite the train's title. I really enjoyed those times and those people. I even took my parents to a Thalassemia Action Group meeting once, so they could see the Foundation I loved in action. I hope people like them continue to be around in the future to help take care of the patients."

I agree. There's a lot more to be done. Problems with patients and families in the recognition of the disease and in its treatment will continue. Advances in diagnosis and treatment will not solve issues like compliance. There will continue to be conflicting personal and cultural attitudes towards new technologies in antenatal diagnosis and new experimental therapies. Patients and their families, and the physicians who care for them, will have to grapple with many current issues as well as many new challenges that these advances in diagnosis and treatment are sure to pose.

Acknowledgments

I have many people to thank. I first want to thank the patients with Cooley's anemia and their families who shared their stories with me and allowed me to tell them in this book. I also want to thank the people at the Cooley's Anemia Foundation, particularly Gina Cioffi and Craig Butler, for their contributions. I am also deeply grateful to my wife, Rona Bank; my sons, Michael and David Bank; Natalie Robins; Una Collins; Vincent Racaniello; Alan Cohen; and Maureen Ward for reading, reviewing and editing the manuscript. Also to Leon Friedman for his always helpful advice as I prepared the manuscript for publication. And, of course, I owe a great debt to all of the students, postdoctoral fellows and colleagues at Columbia University who worked with me in my laboratory over my long career.

Arthur Bank

June 2008

Appendix

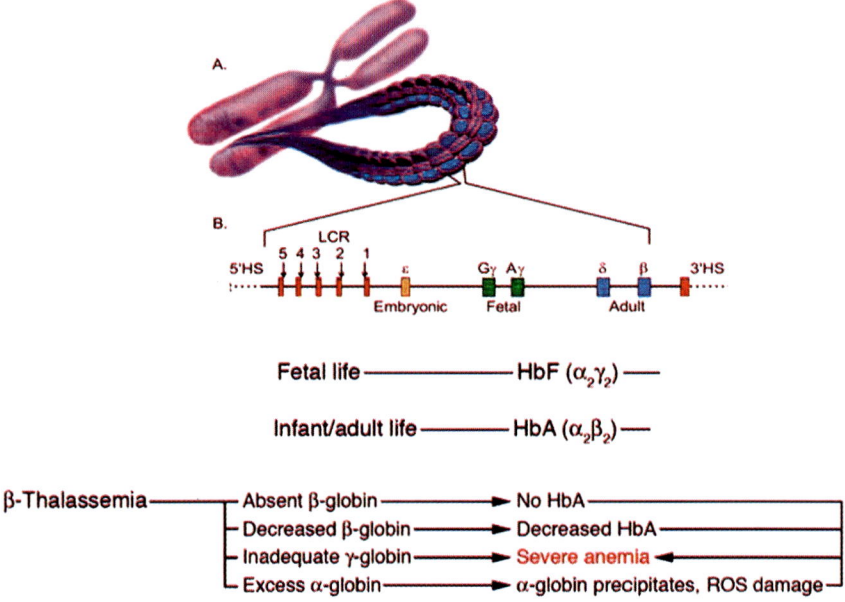

Fig. 1. Summary of normal human hemoglobin production. (A) The human β globin locus (bracketed blue area) is buried in chromosome 11 (in purple). (B) Diagram of the DNA sequences at the human β globin locus. From the left, (5′ end) horizontally are: (1) the β globin locus control region (βLCR) and its hypersensitive sites (HS) 1–5 (in red). The βLCR and its HS are the "enhancers" of the output of all the β globin locus genes, that increase their globin production; (2) the embryonic globin gene (yellow), a minor player, active only in early fetal life; (3) the two fetal globin genes, Gγ and Aγ, the main genes active in fetal life; (4) the human δ globin gene, active but at low levels in adult life; and (5) the human adult β globin gene, the major gene active from birth on. The intergenic γ-δ region is between Aγ and the δ globin genes. To the right of the β globin gene is another HS site (3′). (Adapted from figure originally published in *Blood*. Bank, A. Regulation of human fetal hemoglobin: new players, new complexities. *Blood* 2006; 107:436–443. © The American Society of Hematology).

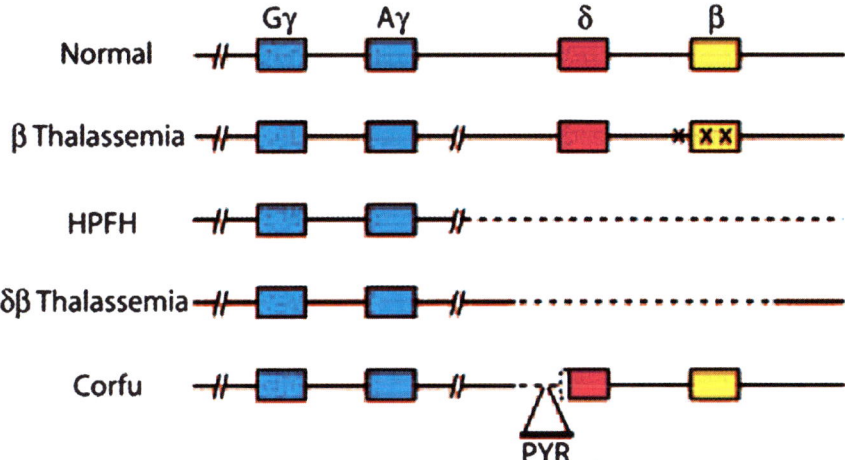

Fig. 2. Mutations and the extent of deletion in thalassemias and HPFH. An X indicates a point mutation. These are most common in β thalassemia. Dotted lines indicate deletions. There is more deletion of γ-δ intergenic sequences in HPFH than in $\delta\beta$ thalassemia. In the naturally occurring Corfu deletion, the PYR sequence and other sequences are deleted upstream of the human δ globin gene, and this γ-δ region deletion is associated with increased human γ globin gene expression (Originally published in *Blood*. Bank A. *Blood*. 2006; 107: 436–443. © The American Society of Hematology).

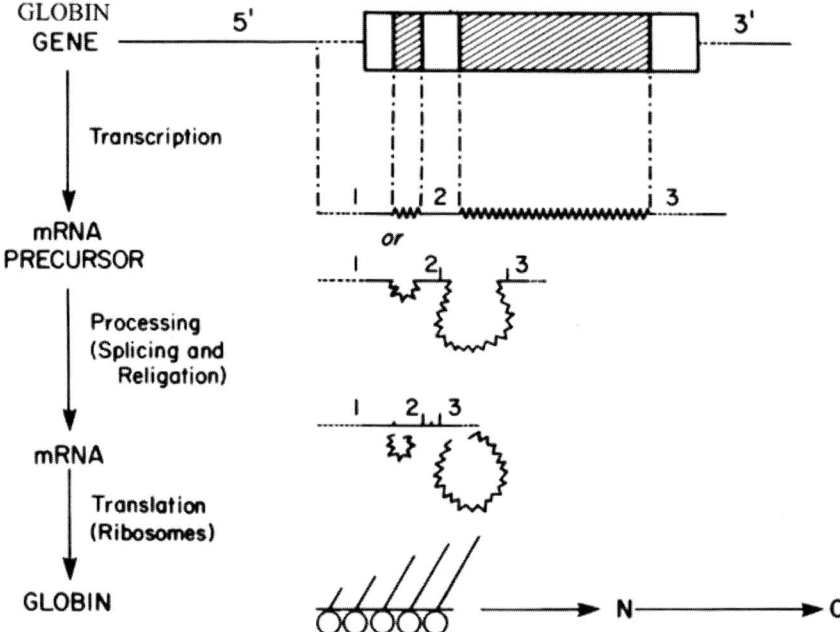

Fig. 3. The processing of intervening sequences. The globin gene DNA coding exon sequences (clear areas) are interrupted by intervening non-coding intron sequences (shaded areas). All of these sequences are transcribed into an mRNA precursor: the horizontal line, with 1, 2, and 3 representing the coding RNA, and the saw-toothed line areas indicating the intron sequences. Splicing and religation of the mRNA precursor result in removal of the introns, and the joining of exons 1, 2 and 3 in the mature mRNA. The mature mRNA is then associated with ribosomes (O) and is translated into globin (N–C).

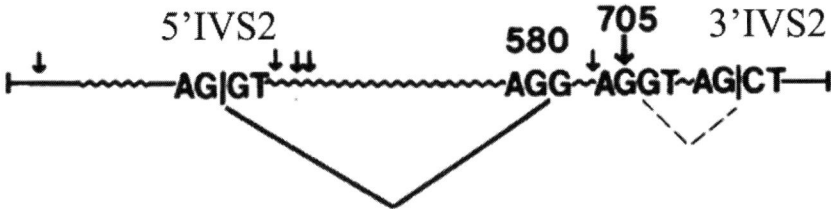

Fig. 4.   The abnormal splicing in IVS2 of Sally Spence's gene. Normally, the splicing of βIVS2 occurs from the AGGT at the 5' end of βIVS2 to the AGCT at the 3' splice end of βIVS2, to remove the entire βIVS2 and permit the formation of mature β globin mRNA that can be translated into β globin. In Sally's gene, two abnormal splices are made: one from the normal 5' end of IVS2 (AGGT) to a cryptic splice site (present in the normal β globin gene) at position 580, (AGG..). This abnormal splice occurs secondary to the thalassemia mutation in Sally's gene at position 705: the 705 mutation creates a new 5' splice site (AGGT) that ties up the normal 3' end of IVS2 (AGCT region), and prevents normal splicing (from the normal 5' end). (This figure was originally published in: Dobkin C, Pergolizzi R, Bahre P, Bank A. Abnormal splice in a mutant human β-globin gene not at the site of a mutation. *Proc. Nat. Acad. Sci.* 1983; 80:1184–1188. Copyright (1983) National Academy of Sciences, U.S.A).

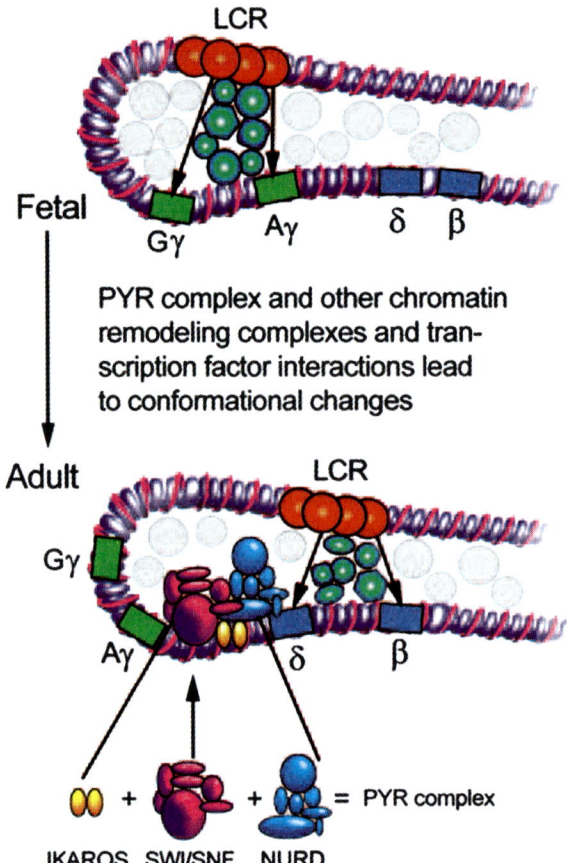

Fig. 5. Human hemoglobin switching. The gray circles indicate unspecified chromatin remodeling complexes and transcription factors. The details of the interactions in chromatin between the βLCR elements, globin structural genes, erythroid transcription factors and these chromatin remodeling complexes are largely unknown. In fetal-embryonic cells, the human βLCR is associated with the γ globin gene. The green circles include the potential activities of fetal stage-specific transcription factors at the LCR and the γ globin promoter. In adult-type cells, interactions of these factors with the βLCR is blocked by PYR complex and other proteins and the βLCR associates with and activates the β globin gene. These interactions lead to repression of γ globin gene expression. PYR complex binding and its HDACs may contribute to this γ globin gene repression. The SWI/SNF complex subunits, the NURD subunits (containing the HDACs), and the DNA binding subunit Ikaros of PYR complex are shown. (Originally published in *Blood*. Bank A. *Blood* 2006; 107:436–443. © The American Society of Hematology).

Selected References

General References

Weatherall DJ, Clegg JB. The Thalassemia Syndromes. Fourth Edition. Blackwell Science, Oxford, England 2001.

Vichinsky E. Cooley's Anemia. Eighth Symposium. Annals of NY Academy of Sciences. Volume 1054, 2005.

Globin Synthesis (Chapter 8)

Weatherall DJ, Clegg JB, Naughton MA. Globin synthesis in thalassaemia: An *in vitro* study. *Nature* 1965; **208**: 1061–1065.

Clegg JB, Naughton MA, Weatherall DJ. An improved method for the characterization of human haemoglobin mutants: Identification of α-2-β-2-95GLU, haemoglobin N (Baltimore). *Nature* 1965; **207**: 945–947.

Bank A, Marks PA. Excess α chain synthesis relative to β chain synthesis in thalassemia major and minor. *Nature* 1966; **212**: 1198–1201.

Bank A, Marks PA. Protein synthesis in a cell free human reticulocyte system: Ribosome function in thalassemia. *J Clin Invest* 1966; **45**: 330–336.

Bank A, Braverman AS, O'Donnell JV, Marks PA. Absolute rates of globin chain synthesis in thalassemia. *Blood* 1968; **31**: 226–233.

Bank A. Hemoglobin synthesis in β-thalassemia: The properties of the free α-chains. *J Clin Invest* 1968; **47**: 860–866.

Braverman AS, Bank A. Changing rates of globin chain synthesis during erythroid cell maturation in thalassemia. *J Mol Biol* 1969; **42**: 57–64.

Bank A, O'Donnell JV. Intracellular loss of free α chains in β thalassemia. *Nature* 1969; **222**: 295–296.

Bank A, O'Donnell JV, Braverman AS. Globin chain synthesis in heterozygotes for β chain mutations. *J Lab Clin Med* 1970; **76**: 616–621.

Kihm AJ, Kong Y, Hong W, *et al*. An abundant erythroid protein that stabilizes free α-hemoglobin. *Nature* 2002; **417**: 758–763.

Feng L, Gell DA, Zhou S, *et al.* Molecular mechanism of AHSP-nucleated stabilization of α hemoglobin. *Cell* 2004; **119**: 629–640.

Kong Y, Zhou S, Kihm AJ, *et al.* Loss of α-hemoglobin stabilizing protein impairs erythropoiesis and excerbates β thalassemia. *J Clin Inv* 2004; **114**: 1457–1466.

Bank A, AHSP: a novel hemoglobin helper. *J Clin Inv* 2007; **117**: 1746–1749.

Globin Messenger RNA (Chapter 9)

Lockand RE and Lingrel JB. The synthesis of mouse hemoglobin β chains in a rabbit reticulocyte system programmed with mouse reticulocyte 9S RNA. *Biochem Biophys Res Comm* 1969; **37**: 204–212.

Nienhuis AW, Anderson WF. Isolation and translation of hemoglobin messenger RNA from thalassemia, sickle cell anemia, and normal human reticulocytes. *J Clin Invest* 1971; **50**: 2458–2460.

Benz EJ, Jr, Forget BG. Defect in messenger RNA for human hemoglobin synthesis in β thalassemia. *J Clin Invest* 1971; **50**: 2755–2760.

Verma IM, Temple GF, Fan H, Baltimore D. *In vitro* synthesis of DNA complementary to rabbit reticulocyte 10S RNA. *Nat New Biol* 1972; **235**: 163–167.

Kacian D, Spiegelman S, Bank A, *et al. In vitro* synthesis of DNA components of human genes for globins. *Nat New Biol* 1972; **235**: 167–169.

Natta C, Banks J, Niazi G, Marks PA, Bank A. Decreased β globin mRNA activity in bone marrow cells in homozygous and heterozygous β thalassemia. *Nat New Biol* 1973; **244**: 280–281.

Kacian DL, Gambino R, Dow LW, *et al.* Decreased globin messenger RNA in thalassemia detected by molecular hybridization. *Proc Natl Acad Sci USA* 1973; **70**: 1886–1890.

Housman D, Forget BG, Skoultchi A, Benz EJ, Jr. Quantitative deficiency of chain-specific globin messenger ribonucleic acids in the thalassemia syndromes. *Proc Natl Acad Sci USA* 1973; **70**: 1809–1813.

Forget BG, Benz EJ, Jr, Skoultchi A, Baglioni C, Housman D. Absence of messenger RNA for β globin chain in β^0 thalassemia. *Nature* 1974; **247**: 379–381.

Kan YW, Holland JP, Dozy AM, Varmus HE. Demonstration of non-functional β-globin mRNA in homozygous β thalassemia. *Proc Natl Acad Sci USA* 1975; **72**: 5140–5144.

Benz EJ, Jr, Swerdlow PS, Forget BG. Absence of functional messenger RNA activity for β globin chain synthesis in β^0-thalassemia. *Blood* 1975; **45**: 1–10.

Globin Genes (Chapters 10 and 11)

Gambino R, Kacian D, O'Donnell J, Ramirez F, Marks PA, Bank A. A limited number of globin genes in human DNA. *Proc Natl Acad Sci USA* 1974; **71**: 3966–3970.

Ramirez F, Natta C, O'Donnell JV, *et al.* Relative numbers of human globin genes assayed with purified α and β complementary human DNA. *Proc Natl Acad Sci USA* 1975; **72**: 1550–1554.

Ottolenghi S, Lanyon WG, Williamson R, Weatherall DJ, Clegg JB, Pitcher CS. Human globin gene analysis for a patient with β-δ β-thalassemia. *Proc Natl Acad Sci USA* 1975; **72**: 2294–2299.

Ramirez F, O'Donnell JV, Natta C, Bank A. Quantitation of human γ globin genes and γ globin mRNA with purified γ globin complementary DNA. *J Clin Invest* 1976; **58**: 1475–1481.

Maniatis T, Hardison RC, Lacy E, *et al.* The isolation of structural genes from libraries of eukaryotic DNA. *Cell* 1978; **15**: 687–701.

Lawn RM, Fritsch EF, Parker RC, Blake G, Maniatis T. The isolation and characterization of linked δ- and β-globin genes from a cloned library of human DNA. *Cell* 1978; **15**: 1157–1174.

Mears JG, Ramirez F, Leibowitz D, *et al.* Changes in restricted human cellular DNA fragments containing globin gene sequences in thalassemia and related disorders. *Proc Natl Acad Sci USA* 1978; **75**: 1222–1226.

Mears JG, Ramirez F, Leibowitz D, Bank A. Organization of human δ and β-globin genes in cellular DNA and the presence of intragenic inserts. *Cell* 1978; **15**: 15–23.

Flavell RA, Kooter JM, De Boer E, Little, PF, Williamson, R. Analysis of the β-δ-globin gene loci in normal and Hb Lepore DNA: Direct determination of gene linkage and intergene distance. *Cell* 1978; **15**: 25–41.

Kan YW, Dozy AM. Polymorphism of DNA sequence adjacent to human β-globin structural gene: relationships to sickle mutation. *Proc. Nat. Acad Sci USA* 1978; **75**: 5631–5635.

Fritsch EF, Lawn RM, Maniatis T. Characterization of deletions which affect the expression of fetal globin genes in man. *Nature* 1979; **279**: 598–603.

Ramirez F, Burns AL, Mears JG, Spence S, Starkman D, Bank A. Isolation and characterization of cloned human fetal globin genes. *Nucleic Acids Res* 1979; **7**: 1147–1162.

Orkin SH, Old JM, Weatherall DJ, Nathan DG. Partial deletion of β-globin gene DNA in certain patients with β^0-thalassemia. *Proc Natl Acad Sci USA* 1979; **76**: 2400–2404.

Chang JC, Kan YW. Beta zero thalassemia, a nonsense mutation in man. *Proc Natl Acad Sci USA* 1979; **76**: 2886–2889.

Lawn RM, Efstratiadis A, O'Connell C, Maniatis T. The nucleotide sequence of the human β-globin gene. *Cell* 1980; **21**: 647–651.

Spritz RA, Jagadeeswaran P, Choudary PV, *et al.* Base substitution in an intervening sequence of a β^+-thalassemic human globin gene. *Proc Natl Acad Sci USA* 1981; **78**: 2455–2459.

Pergolizzi R, Spritz RA, Spence S, Goossens M, Kan YW, Bank A. Two cloned β thalassemia genes are associated with amber mutations at codon 39. *Nucleic Acids Res* 1981; **9**: 7065–7072.

Busslinger M, Moschonas N, Flavell RA. β^+ thalassemia: Aberrant splicing results from a single point mutation in an intron. *Cell* 1981; **27**: 289–298.

Burns AL, Spence S, Kosche K, *et al*. Isolation and characterization of cloned DNA: the δ and β globin genes in homozygous β^+ thalassemia. *Blood* 1981; **57**: 140–146.

Baird M, Driscoll C, Schreiner H, *et al*. A nucleotide change at a splice junction in the human β-globin gene is associated with β^0-thalassemia. *Proc Natl Acad Sci USA* 1981; **78**: 4218–4221.

Antonarakis SE, Boehm CD, Giardina PJ, Kazazian HH Jr. Non-random association of polymorphic restriction sites in the β-globin gene cluster. *Proc Nat Acad Sci USA* 1982; **10**: 1283–1294.

Spence SE, Pergolizzi RG, Donovan-Peluso M, Kosche KA, Dobkin CS, Bank A. Five nucleotide changes in the large intervening sequence of a β globin gene in a β^+ thalassemia patient. *Nucleic Acids Res* 1982; **10**: 1283–1294.

Treisman R, Orkin SH, Maniatis T. Specific transcription and RNA splicing defects in five cloned β-thalassaemia genes. *Nature* 1983; **302**: 591–596.

Dobkin C, Pergolizzi RG, Bahre P, Bank A. Abnormal splice in a mutant human β-globin gene not at the site of a mutation. *Proc Natl Acad Sci USA* 1983; **80**: 1184–1188.

Donovan-Peluso M, Young K, Dobkin C, Bank A. Erythroleukemia (K562) cells contain a functional β-globin gene. *Mol Cell Biol* 1984; **4**: 2553–2555.

Dobkin C, Bank A. Reversibility of IVS 2 missplicing in a mutant human β-globin gene. *J Biol Chem* 1985; **260**: 16332–16337.

Atweh GF, Anagnou NP, Shearin J, Forget BG, Kaufman RE. β-thalassemia resulting from a single nucleotide substitution in an acceptor splice site. *Nucleic Acids Res* 1985; **13**: 777–790.

Metherall JE, Collins FS, Pan J, Weissman SM, Forget BG. β^0 thalassemia caused by a base substitution that creates an alternative splice acceptor site in an intron. *Embo J* 1986; **5**: 2551–2557.

Atweh GF, Forget BG. Identification of a β-thalassemia mutation associated with a novel haplotype of RFLPs. *Am J Hum Genet* 1986; **38**: 855–859.

Stoeckert CJ, Jr, Metherall JE, Yamakawa M, Eisenstadt JM, Weissman SM, Forget BG. Expression of the affected A γ globin gene associated with Greek nondeletion hereditary persistence of fetal hemoglobin. *Mol Cell Biol* 1987; **7**: 2999–3003.

Lapoumeroulie C, Acuto S, Rouabhi F, Labie D, Krishnamoorthy R, Bank A. Expression of a β thalassemia gene with abnormal splicing. *Nucleic Acids Res* 1987; **15**: 8195–8204.

LaFlamme S, Acuto S, Markowitz D, Vick L, Landschultz W, Bank A. Expression of chimeric human β- and δ-globin genes during erythroid differentiation. *J Biol Chem* 1987; **262**: 4819–4826.

Donovan-Peluso M, Acuto S, Swanson M, Dobkin C, Bank A. Expression of human γ-globin genes in human erythroleukemia (K562) cells. *J Biol Chem* 1987; **262**: 17051–17057.

Atweh GF, Wong C, Reed R, *et al*. A new mutation in IVS-1 of the human β globin gene causing β thalassemia due to abnormal splicing. *Blood* 1987; **70**: 147–151.

Fetal Hemoglobin (Chapter 12)

Bank A, Mears JG, Ramirez F. Disorders of human hemoglobin. *Science* 1980; **207**: 486–493.

Groudine M, Peretz M, Weintraub H. Transcriptional regulation of hemoglobin switching in chicken embryos. *Mol Cell Biol* 1981; **1**: 281–288.

Forget BG, Tuan D, Newman MV, *et al.* Molecular studies of mutations that increase Hb F production in man. *Prog Clin Biol Res* 1983; **134**: 65–76.

Tuan D, Feingold E, Newman M, Weissman SM, Forget BG. Different 3' end points of deletions causing δ β-thalassemia and hereditary persistence of fetal hemoglobin: implications for the control of γ-globin gene expression in man. *Proc Natl Acad Sci USA* 1983; **80**: 6937–6941.

Anagnou NP, Moulton AD, Keller G, *et al.* Cis-acting sequences that affect the expression of the human fetal γ-globin genes. *Prog Clin Biol Res* 1985; **191**: 163–182.

Tuan D, Solomon W, Li Q, London IM. The "β-like globin" gene domain in human erythroid cells. *Proc Natl Acad Sci*, USA 1985; **82**: 6384–88.

Forrester WC, Thompson C, Elder JT, Groudine M. A developmentally stable chromatin structure in the human β-globin gene cluster. *Proc Natl Acad Sci USA* 1986; **83**: 1359–1363.

Grosveld F, van Assendelft GB, Greaves DR, Kollias G. Position-independent, high-level expression of the human β-globin gene in transgenic mice. *Cell* 1987; **51**: 975–985.

Feingold EA, Forget BG. The breakpoint of a large deletion causing hereditary persistence of fetal hemoglobin occurs within an erythroid DNA domain remote from the β-globin gene cluster. *Blood* 1989; **74**: 2178–2186.

Gallarda JL, Foley KP, Yang ZY, Engel JD. The β-globin stage selector element factor is erythroid-specific promoter/enhancer binding protein NF-E4. *Genes Dev* 1989; **3**: 1845–1859.

Talbot D, Collis P, Antoniou M, Vidal M, Grosveld F, Greaves DR. A dominant control region from the human β-globin locus conferring integration site-independent gene expression. *Nature* 1989; **338**: 352–355.

Galanello R, Melis MA, Podda A, *et al.* Deletion δ-thalassemia: The 7.2 kb deletion of Corfu δ β-thalassemia in a non-β-thalassemia chromosome. *Blood* 1990; **75**: 1747–1749.

Enver T, Raich N, Ebens AJ, *et al.* Developmental regulation of human fetal-to-adult globin gene switching in transgenic mice. *Nature* 1990; **344**: 309–313.

O'Neill D, Bornschlegel K, Flamm M, Castle M, Bank A. A DNA-binding factor in adult hematopoietic cells interacts with a pyrimidine-rich domain upstream from the human δ-globin gene. *Proc Natl Acad Sci USA* 1991; **88**: 8953–8957.

Jane SM, Ney PA, Vanin EF, Gumucio DL, Nienhuis AW. Identification of a stage selector element in the human γ-globin gene promoter that fosters preferential interaction with the 5' HS2 enhancer when in competition with the β-promoter. *Embo J* 1992; **11**: 2961–2969.

Fraser P, Pruzina S, Antoniou M, Grosveld F. Each hypersensitive site of the human β-globin locus control region confers a different developmental pattern of expression on the globin genes. *Genes Dev* 1993; **7**: 106–113.

Tjian R, Maniatis T. Transcriptional activation a complex puzzle with few easy pieces. *Cell* 1994; **77**: 5–8.

Anagnou NP, Perez-Stable C, Gelinas R, *et al.* Sequences located 3' to the breakpoint of the hereditary persistence of fetal hemoglobin-3 deletion exhibit enhancer activity and can modify the developmental expression of the human fetal A γ-globin gene in transgenic mice. *J Biol Chem* 1995; **270**: 10256–10263.

Thanos D, Maniatis T. Viral induction of human IFN beta gene expression requires the assembly of an enhanceosome. *Cell* 1995; **83**: 1091–1100.

Bresnick EH, Tze L. Synergism between hypersensitive sites confers long-range gene activation by the β-globin locus control region. *Proc Natl Acad Sci USA* 1997; **94**: 4566–4571.

O'Neill D, Yang J, Erdjument-Bromage H, Bornschlegel K, Tempst P, Bank A. Tissue-specific and developmental stage-specific DNA binding by a mammalian SWI/SNF complex associated with human fetal-to-adult globin gene switching. *Proc Natl Acad Sci USA* 1999; **96**: 349–354.

Kim J, Sif S, Jones B, Jackson A, Koipally J, *et al.* Ikaros DNA-binding proteins direct formation of chromatin remodeling complexes in lymphocytes. *Immunity* 1999; **10**: 345–55.

Bender MA, Bulger M, Close J, Groudine M. β-globin gene switching and DNase I sensitivity of the endogenous β-globin locus in mice do not require the locus control region. *Mol Cell* 2000; **5**: 387–393.

Forsberg EC, Downs KM, Bresnick EH. Direct interaction of NF-E2 with hypersensitive site 2 of the β-globin locus control region in living cells. *Blood* 2000; **96**: 334–339.

Forsberg EC, Downs KM, Christensen HM, Im H, Nuzzi PA, Bresnick EH. Developmentally dynamic histone acetylation pattern of a tissue-specific chromatin domain. *Proc Natl Acad Sci USA* 2000; **97**: 14494–14499.

O'Neill DW, Schoetz SS, Lopez RA, *et al.* An Ikaros-containing chromatin-remodeling complex in adult-type erythroid cells. *Mol Cell Biol* 2000; **20**: 7572–7582.

Bulger M, Sawado T, Schubeler D, Groudine M. ChIPs of the β-globin locus: unravelling gene regulation within an active domain. *Curr Opin Genet Dev* 2002; **12**: 170–177.

Lopez RA, Schoetz S, DeAngelis K, O'Neill D, Bank A. Multiple hematopoietic defects and delayed globin switching in Ikaros null mice. *Proc Natl Acad Sci USA* 2002; **99**: 602–607.

Tanabe O, Katsuoka F, Campbell AD, *et al.* An embryonic/fetal β-type globin gene repressor contains a nuclear receptor TR2/TR4 heterodimer. *Embo J* 2002; **21**: 3434–3442.

Tolhuis B, Palstra RJ, Splinter E, Grosveld F, de Latt W. Looping and interaction between hypersensitive sites in the active β-globin locus. *Mol Cell* 2002; **10**: 1453–1465.

Palstra RJ, Tolhuis B, Splinter E, Nijmeijer R, Grosveld F, de Laat W. The β-globin nuclear compartment in development and erythroid differentiation. *Nat Genet* 2003; **35**: 190–194.

Drissen R, Palstra RJ, Gillemans N, *et al.* The active spatial organization of the β-globin locus requires the transcription factor EKLF. *Genes Dev* 2004; **18**: 2485–2490.

Patrinos GP, de Krom M, de Boer E, *et al.* Multiple interactions between regulatory regions are required to stabilize an active chromatin hub. *Genes Dev* 2004; **18**: 1495–1509.

Zhou W, Zhao Q, Sutton R, *et al.* The role of p22 NF-E4 in human globin gene switching. *J Biol Chem* 2004; **279**: 26227–26232.

Omori A, Tanabe O, Engel JD, Fukamizu A, Tanimoto K. Adult stage gamma-globin silencing is mediated by a promoter direct repeat element. *Mol Cell Biol* 2005; **25**: 3443–3451.

Bank A. Understanding globin regulation in β-thalassemia: it's as simple as α, β, γ, δ. *J Clin Invest* 2005; **115**: 1470–1473.

Chakalova L, Osborne CS, Dai YF, *et al.* The Corfu δ β-thalassemia deletion disrupts γ-globin gene silencing and reveals post-transcriptional regulation of HbF expression. *Blood* 2005; **105**: 2154–2160.

Vakoc CR, Letting DL, Gheldof N, *et al.* Proximity among distant regulatory elements at the β-globin locus requires GATA-1 and FOG-1. *Mol Cell* 2005; **17**: 453–462.

Bank A. Regulation of human fetal hemoglobin: new players, new complexities. *Blood* 2006; **107**: 435–443.

Zhao Q, Zhou W, Rank G, *et al.* Repression of human γ-globin gene expression by a short isoform of the NF-E4 protein is associated with loss of NF-E2 and RNA polymerase II recruitment to the promoter. *Blood* 2006; **107**: 2138–2145.

Keys A, Tallack MR, Zhan Y, *et al.* A mechanism for Ikaros regulation in human globin gene switching. *Br J Hem* 2008; **141**: 398–406

Globin Gene Therapy (Chapter 15)

Young K, Donovan-Peluso M, Bloom K, Allan M, Paul J, Bank A. Stable transfer and expression of exogenous human globin genes in human erythroleukemia (K562) cells. *Proc Natl Acad Sci USA* 1984; **81**: 5315–5319.

Bank A, Donovan-Peluso M, Lerner N, Rund D. Human globin gene expression after gene transfer. *Blood Cells* 1987; **13**: 269–275.

Markowitz D, Goff S, Bank A. Construction and use of a safe and efficient amphotropic packaging cell line. *Virology* 1988; **167**: 400–406.

Markowitz D, Goff S, Bank A. A safe packaging line for gene transfer: separating viral genes on two different plasmids. *J Virol* 1988; **62**: 1120–1124.

Hesdorffer C, Ward M, Markowitz D, Bank A. Efficient gene transfer in live mice using a unique retroviral packaging line. *DNA Cell Biol* 1990; **9**: 717–723.

Hesdorffer C, Antman K, Bank A, Fetell M, Mears G, Begg M. Human MDR gene transfer in patients with advanced cancer. *Hum Gene Ther* 1994; **5**: 1151–1160.

Leboulch P, Huang GM, Humphries RK, *et al.* Mutagenesis of retroviral vectors transducing human β-globin gene and β-globin locus control region derivatives results in stable transmission of an active transcriptional structure. *Embo J* 1994; **13**: 3065–3076.

Ward M, Richardson C, Pioli P, *et al.* Transfer and expression of the human multiple drug resistance gene in human CD34+ cells. *Blood* 1994; **84**: 1408–1414.

Sadelain M, Wang CH, Antoniou M, Grosveld F, Mulligan RC. Generation of a high-titer retroviral vector capable of expressing high levels of the human β-globin gene. *Proc Natl Acad Sci USA* 1995; **92**: 6728–6732.

Ward M, Pioli P, Ayello J, *et al.* Retroviral transfer and expression of the human multiple drug resistance (MDR) gene in peripheral blood progenitor cells. *Clin Cancer Res* 1996; **2**: 873–876.

Gallardo HF, Tan C, Ory D, Sadelain M. Recombinant retroviruses pseudotyped with the vesicular stomatitis virus G glycoprotein mediate both stable gene transfer and pseudotransduction in human peripheral blood lymphocytes. *Blood* 1997; **90**: 952–957.

Raftopoulos H, Ward M, Leboulch P, Bank A. Long-term transfer and expression of the human β-globin gene in a mouse transplant model. *Blood* 1997; **90**: 3414–3422.

Hesdorffer C, Ayello J, Ward M, *et al.* Phase I trial of retroviral-mediated transfer of the human MDR1 gene as marrow chemoprotection in patients undergoing high-dose chemotherapy and autologous stem-cell transplantation. *J Clin Oncol* 1998; **16**: 165–172.

Abonour R, Williams DA, Einhorn L, *et al.* Efficient retrovirus-mediated transfer of the multidrug resistance 1 gene into autologous human long-term repopulating hematopoietic stem cells. *Nat Med* 2000; **6**: 652–658.

May C, Rivella S, Callegari J, *et al.* Therapeutic haemoglobin synthesis in β-thalassaemic mice expressing lentivirus-encoded human β-globin. *Nature* 2000; **406**: 82–86.

Cavazzana-Calvo M, Hacien-Bey S, de Saint Basile G, *et al.* Gene therapy of human severe combined immunodeficiency (SCID)-XI disease. *Science* 2000; **288**: 669–672.

Pawliuk R, Westerman KA, Fabry ME, *et al.* Correction of sickle cell disease in transgenic mouse models by gene therapy. *Science* 2001; **294**: 2368–2371.

Bradley MB, Sattler RM, Raftopoulos H, *et al.* Correction of phenotype in a thalassemia mouse model using a nonmyeloablative marrow transplantation regimen. *Biol Blood Marrow Transplant* 2002; **8**: 453–461.

Imren S, Payen E, Westerman KA, *et al.* Permanent and panerythroid correction of murine β-thalassemia by multiple lentiviral integration in hematopoietic stem cells. *Proc Natl Acad Sci USA* 2002; **99**: 14380–14385.

May C, Rivella S, Chadburn A, Sadelain M. Successful treatment of murine β-thalassemia intermedia by transfer of the human β-globin gene. *Blood* 2002; **99**: 1902–1908.

Hacein-Bey-Abina S, Le Deist F, Carlier F, *et al.* Sustained correction of X-linked severe combined immunodeficiency by ex vivo gene therapy. *N Engl J Med* 2002; **346**: 1185–1193.

Rivella S, May C, Chadburn A, Riviere I, Sadelain M. A novel murine model of Cooley anemia and its rescue by lentiviral-mediated human β-globin gene transfer. *Blood* 2003; **101**: 2932–2939.

Imren S, Fabry ME, Westerman KA, *et al.* High-level β-globin expression and preferred intragenic integration after lentiviral transduction of human cord blood stem cells. *J Clin Invest* 2004; **114**: 953–962.

Bank A, Dorazio R, Leboulch P. A phase I/II clinical trial of β-globin gene therapy for β-thalassemia. *Ann N Y Acad Sci* 2005; **1054**: 308–316.

Takahashi K, Yamanaka S. Induction of pluripotent stem cells from mouse embryonic and adult fibroblast cultures by defined factors. *Cell* 2006; **126**: 663–676.

Cavazzana-Calvo M, Fischer A. Gene therapy for severe combined immuno-deficiency: are we there yet? *J Clin Invest* 2007; **117**: 1456–1465.

Takahashi K, Tanabe K, Ohnuki M, *et al.* Induction of pluripotent stem cells from adult human fibroblasts by defined factors. *Cell* 2007; **131**: 861–872.

Hanna J, Wernig M, Markoulaki S, *et al.* Treatment of sickle cell anemia mouse model with iPS Cells Generated from autologous skin. *Science* 2007; **318**: 1920–1923

Jaenisch R, Young R. Stem cells, the molecular circuitry of pluripotency and nuclear reprogramming. *Cell* 2008; **132**: 567–582

About the Author

Dr. Arthur Bank has been a scientist, clinician and teacher at Columbia University and Presbyterian Hospital for over 40 years. He was Head of the Division of Hematology at Columbia-Presbyterian Medical Center. He has been a leader in Cooley's anemia research, has published extensively on the mechanisms of human globin gene regulation and human gene therapy, and has been continuously funded by grants from the National Institutes of Health throughout his career. Dr. Bank is currently Professor Emeritus of Medicine and of Genetics and Development at Columbia University. A graduate of Columbia College and Harvard Medical School, and a nationally and internationally recognized authority in the field of hematology research, Dr. Bank has written a book about a blood disease he has studied extensively, Cooley's anemia.

Index

Naughton, Michael, 92
Negative iron balance, 12, 185, 200,
    202, 241
Neutral polymorphisms, 149
New York Academy of Sciences, 41
New York Daily News, 25
New York Hospital (NYH), 17, 18, 22,
    24, 25, 27, 33, 62, 67, 77–79, 82, 110,
    197, 198, 201, 241
New York Presbyterian Hospital, 18,
    185, 198
Newly synthesized globin, 93, 95, 101
NF-E2, 162, 178
Nienhuis, Arthur, 108
NIH. *See* National Institutes of Health
Non-compliance, 200
Nonsense mutation, 155
Normal adult human hemoglobin
    (hemoglobin A, HbA), 113, 191
Normal β globin DNA, 5, 6, 13, 14, 98,
    125, 208
Normal red blood cells, 4, 10, 89, 91,
    100, 156, 162, 190
Novartis, 68, 186, 188, 189
Nuclear extracts, 166, 168, 176
Nucleated red blood cells, 3, 9, 13, 91,
    99, 105, 168, 191
NURD, 173, 256
NYH, 22, 197, 198

Oct4, 215
O'Donnell, Joyce, 97, 101, 115
OligodT, 110
Oligonucleotide hybridization for
    prenatal diagnosis, 226
Oligonucleotides, 166, 167, 172, 226
Olivieri, Nancy, 188, 231
O'Neill, David, 166, 170–172, 174–176
Oral iron chelators, 61
    Exjade, 13, 68, 83, 187, 189, 202,
        228, 231, 242, 248
    L1, 13, 68, 187–190, 202, 230, 231,
        242, 248
O'Reilly, Richard, 201–202
Orkin, Stuart, 148–149

Oxygen, 2–8, 10, 11, 13, 37, 50, 78, 91,
    159
    in hemoglobin, 2, 3
    in red blood cells, 2
    oxygen affinity, 8, 159
Oxygen affinity of hemoglobin, 8, 159
Oxygenated hemoglobin, 3

Pacemaker. *See* cardiac pacemaker
Papayannopoulou, Thalia, 49
Paradiso (family)
    Paradiso, Connie, 52–61, 243, 247
    Paradiso, Edward, 53, 54, 60
    Paradiso, Janice, 55, 60
    Paradiso, Paul, 55–57
    Paradiso, Peter, 53–55, 58, 60
    Paradiso, Susan, 53, 54
Paris trial, 212, 214
Pergolizzi, Robert, 144, 255
Pericarditis, 30, 39
Perkins, Andrew, 179
Phase 1 clinical trial, 210
Piomelli, Sergio, 40, 234, 246
Pizzulli, Amelia (Amy), 74–84, 186,
    200, 242
Pizzulli, Danny, 79–81, 83
Placenta, 8, 159
Plastic (Plstc) mice, 179
Polymerase chain reaction (PCR), 212
Polymorphisms in DNA, 149
Polyribosome-mRNA complex, 91
Prenatal diagnosis. *See* antenatal
    diagnosis
Propper, Richard, 186
Protein. *See* globin, hemoglobin
Protein complexes, 163, 164, 166
    NURD, 173, 256
    PYR complex, 159, 163, 164, 167,
        181, 256
    SWI/SNF, 172–174, 178, 256
Protein-DNA complexes, 166
Protein synthesis, 88–90, 92, 105, 136
Proteolysis, 100
PYR complex, 159, 163, 164, 167–174,
    176–181, 256